In *Flo...* Sandra... Each cha... introduces a single flower during a single hour, highlighting twenty-four different species from around the world. Beginning at midnight, we spot the long tubular flowers of the moonflower, *Ipomoea alba*; they attract a frenzy of hawk moths before dawn arrives, and the flowers wither and collapse. As day breaks, dandelions and chicory open their heads, supplying flies and bees with the energy they need to lay eggs and raise their young. That evening, we smell night-blooming jessamine's powerful—some may say nauseating—sweet scent.

For each hour in our flower day, celebrated artist Katie Scott has depicted these scenes with gorgeous pen and ink illustrations. Working closely together to narrate and illustrate these unique moments in time, Knapp and Scott have created an engaging read that is a perfect way to spend an hour or two—and a true gift for amateur botanists, gardeners, and anyone who wants to stop to appreciate the flowers.

Praise for **FLOWER DAY**

"All of us are used to walking out into gardens in the evening and seeing the moonflowers opening, with Carolina jessamine and others joining them throughout the night. Buttercups, asters, sunflowers, and fuchsias bloom at specific times, starting near dawn. Together, these plants present a rich panorama—a kind of floral clock that, with its various combinations, informs a careful observer of the time of day, rain or shine. In this beautifully illustrated book, Knapp generously enriches our understanding of the lovely floral views that greet us every hour of the day and night."

PETER H. RAVEN, president emeritus,
Missouri Botanical Garden

"Knapp's extraordinary book will inspire every reader to look more closely at the world's botanical wonders. The flowers she introduces abound with a diversity of color, shape, odor, height, size, sex habits, pollinators, fruits, seeds, patterns, and longevity. From the enormous sago palm to the tiny hair grass, and from our lust for orchids to our love-hate relationship with dandelions, these stories are beautifully written and illustrated. Every reader will come away with a greater admiration for all things floral."

MEG LOWMAN, author of *The Arbornaut*

"*Flower Day* offers a glorious, round-the-clock glimpse into the incredible plant diversity on the planet. This book highlights how and why different plants flower at different times during the day, and Knapp masterfully intertwines the botany, plant biology, history, evolution, and ecology of twenty-four species. Anyone with even a cursory affinity for plants or curiosity about the natural world will appreciate this beautiful book."

ALLISON MILLER, Donald Danforth Plant Science Center and Saint Louis University

Flower Day

FLOWER DAY

A STORY OF 24 HOURS AND
24 FLORAL LIVES

WRITTEN BY	ILLUSTRATED BY
Sandra Knapp	Katie Scott

The University of Chicago Press
Chicago and London

The University of Chicago Press, Chicago 60637
The University of Chicago Press, Ltd., London
© 2025 by Sandra Knapp
Illustrations © 2025 by Katie Scott
All rights reserved. No part of this book may be used or reproduced in any manner whatsoever without written permission, except in the case of brief quotations in critical articles and reviews. For more information, contact the University of Chicago Press, 1427 E. 60th St., Chicago, IL 60637.
Published 2025
Printed in China

34 33 32 31 30 29 28 27 26 25   1 2 3 4 5

ISBN-13: 978-0-226-83452-8 (cloth)
ISBN-13: 978-0-226-83453-5 (e-book)
DOI: https://doi.org/10.7208/chicago/9780226834535.001.0001

Library of Congress Cataloging-in-Publication Data

Names: Knapp, Sandra, author. | Scott, Katie, illustrator.
Title: Flower day : a story of 24 hours and 24 floral lives / written by
    Sandra Knapp ; illustrated by Katie Scott
Other titles: Earth day (University of Chicago. Press)
Description: Chicago : The University of Chicago Press, 2025. | Series:
    Earth day | Includes bibliographical references and index.
Identifiers: LCCN 2024047447 | ISBN 9780226834528 (cloth)
    | ISBN 9780226834535 (ebook)
Subjects: LCSH: Flowers. | Flowers—Pictorial works.
Classification: LCC QK653 .K5 2025 | DDC 582.13—dc23/eng/20241115
LC record available at https://lccn.loc.gov/2024047447

♾This paper meets the requirements of ANSI/NISO Z39.48-1992
(Permanence of Paper).

*For my
beloved granddaughters,
Libby and Julia—
may their days be filled
with flowers.*

Contents

PREFACE XI

MIDNIGHT
Moonflower
(THE AMERICAS)
3

1AM
Queen of the Night
(THE CARIBBEAN,
CENTRAL AMERICA,
MEXICO)
9

2AM
Angraecum
(MADAGASCAR)
15

3AM
East Coast Banksia
(AUSTRALIA)
21

4AM
Chicory
(EUROPE, INTRODUCED
TO NORTH AMERICA)
27

5AM
Dandelion
(WORLDWIDE)
33

6 AM
Coyote Tobacco
(NORTH AMERICA)
39

7 AM
Curly Rock Rose
(AFRICA, EUROPE)
47

8 AM
Uape Jacana or Giant Waterlily
(AMAZON BASIN, THE GUIANAS)
53

9 AM
Sunflower
(NORTH AMERICA)
59

10 AM
Sago Palm
(SOUTHEAST ASIA)
65

11 AM
Western Blue Flax or Lewis Flax
(NORTH AMERICA)
71

NOON
Jack-Go-to-Bed-at-Noon or Meadow Salsify
(EUROPE, INTRODUCED TO NORTH AMERICA)
77

1 PM
Antarctic Hair Grass
(ANTARCTICA, PATAGONIA)
83

2 PM
Titan Arum or Corpse Flower
(SUMATRA)
89

3 PM
The Traveler's Tree
(MADAGASCAR)
97

4 PM
Four-O'Clock
(THE AMERICAS)
103

5 PM
Harakeke or New Zealand Flax
(AOTEAROA—NEW ZEALAND)
109

6 PM
California Poppy
(NORTH AMERICA)
115

7 PM
Night-Blooming Jessamine
(THE CARIBBEAN)
123

8 PM
White Campion
(EURASIA, INTRODUCED TO NORTH AMERICA)
129

9 PM
Sacred Datura
(NORTH AMERICA)
137

10PM
Leafless Ephedra
(EUROPE)
143

11PM
Giant Saguaro
(NORTH AMERICA)
149

Epilogue
155

ACKNOWLEDGMENTS 159

FURTHER READING 161   INDEX 175

Preface

Georgia O'Keeffe, the artist whose extraordinary paintings surely inspired my love of flowers, said, "When you take a flower in your hand and really look at it, it's your world for the moment." A flower is a world of interactions and happenings.

The parts of flowers that we see and appreciate are the tepals—the usually leafy sepals and often brightly colored petals. Petals come in an astonishing variety of shapes and sizes, from tiny green nubbins in grasses to flashy magenta tubes in four-o'clocks to large floppy white straps in night-flowering cacti. The showy petals of our favorite flowers, however much we might enjoy them, are not there for us. Flowers must attract pollinators—a cast of characters including birds, bees, and even lemurs—who carry pollen, ensuring future generations of plants.

If you want to enjoy flowers in a pot or in a garden, you might plant seeds. But where do seeds come from? Let's talk about plant reproduction, the birds and the bees. Seeds develop inside fruits—this, by the way, is why a tomato is a fruit, not a vegetable! For seeds to form, pollen and ovules must meet. Where, then, does pollen come from? Plants make pollen in their anthers, which come in shapes and sizes as varied as those of petals. Pollen grains are tiny, like dust particles— think of the lilies on your kitchen table whose dark-brown pollen sheds all over everything when the anthers open. When anthers are ready for action, they open at a slit or tiny pore and release the pollen grains. Anthers usually sit atop long stalks called filaments; together they are stamens.

And where do ovules come from? Ovules grow inside a flower's ovary, and it is this fact that gives flowering plants their scientific name—angiosperms, meaning "hidden seeds." Variation is the watchword of most things associated with flowers, but pistils are often shaped like a long-necked bottle: its body is the ovary, its contents are the ovules, and its long neck is the style. The style is the path along which pollen must travel from the outside world to the ovary and ovules. At the tip of the style is the stigma. The stigma is often sticky, helping it catch passing pollen.

So pollen meets ovule, and a seed begins to grow. This is sexual reproduction, but some plants, like the Antarctic hair grass we will meet around midday, can reproduce without sex. Vegetative reproduction can occur in many forms: the hair grass extends

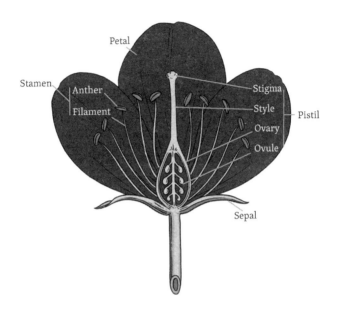

stems along the ground, making new plantlets as it goes; the potato grows new plants from the "eyes" of its tubers; and other plants, like the dandelion, even produce seeds without pollen and ovule meeting, in a phenomenon called apomixis. But plant sex and the ways in which these apparently static organisms set pollen and ovule up are truly mind-boggling. People who study animals are always astounded at the myriad ways plants combine their sexual functions. Many flowers are cosexual; they bear both stamens and pistils. For some, like the blue flax, these structures are out in the open for all to see; for others, like the traveler's tree, they are hidden away inside the flower. Other species have some flowers with stamens but no pistils and others with pistils but no stamens. When both staminate and pistillate flowers are on the same individual, the plant is called monoecious ("one house"), but when they're on different individuals, the plant—like the white campion we will meet near the end of the day—is called dioecious ("two houses"). It doesn't stop there:

Some plants have some flowers with both stamens and pistils and others with only pistils or only stamens. And sometimes the flowers of dioecious plants look so different that it's hard to match the individuals up. I have spent many hours hunting plants of one type or the other during my trips collecting plants in the rainforests of the tropics.

Why all this variation? "Pollen meets ovule" seems straightforward, doesn't it? But plants must rely on some mechanism to get pollen from one flower to another, or even from the stamen to the stigma; after all, plants don't move like animals do to find mates. Why don't flowers simply drop their pollen on their own stigma? Some, like those of the night-blooming jessamine we'll meet around sunset, do just that. Self-fertilization comes with a hitch, though. Natural selection (the process through which organisms best adapted to their environments survive and reproduce) favors genetic variation, and outcrossing (when pollen from one plant meets the ovule of another) promotes genetic variation by combining genes from two parents. Self-fertilizing plants don't make seeds with as much genetic diversity and may therefore face steeper odds in surviving and passing their genes on to the next generation, especially if conditions change. Plants encourage outcrossing in lots of ways, including by altering the timing of anther opening and stigma receptivity so that the pollen of one flower never gets onto the stigma of the same flower. Banksias, for example, hide their sticky stigma away until very late in flowering. The Amazonian giant waterlily is even more extreme, presenting its sticky stigmas one night, closing its flowers during the day, then opening its anthers the next evening to release pollen. Dioecious plants, those with pollen on one individual and ovules on another, can do nothing but outcross. But many other plants, despite significant efforts to get pollen from one individual to another, still self-fertilize as a last resort. Better to have some seeds than none at all.

So, plants can encourage outcrossing, but what actually moves pollen from one flower to another? Sometimes it's the wind, as for our Antarctic hair grass or the oak trees whose pollen covers cars parked on the street in the spring, but often this is where the birds and the bees—and many other animals—come in. Animals carry pollen, usually unwittingly, from one flower to another. These are pollinators, and like the flowers themselves, they are astonishingly diverse. Bats seek out the blousy flowers of the queen of the night, whose name reflects the time its flowers open for action; lemurs carry pollen from one traveler's tree to another; and many types of insects, such as flies, crickets, wasps, and of course bees, get covered with pollen when they land on a flower, then bring the grains with them to the next. Why do all these animals visit flowers, though? What's in it for them?

To attract visitors, flowers offer rewards—sometimes the pollen itself, which is high in nitrogen and very nutritious for insects like bees, who use the pollen they collect from California poppies to feed their growing larvae. Other flowers, like those of the saguaro cactus, are filled with sweet nectar, sought after by birds and mammals alike. Others still, like the Amazonian giant waterlily, have flower parts specifically for pollinators to eat. What's important is that a pollinator is sufficiently rewarded so that, once attracted to a flower, it goes on to another of the same type—thereby acting as matchmaker for pollen and ovule.

In the eighteenth century, the great German botanist and playwright Johann Wolfgang von Goethe first suggested that flowers were leaves for reproduction. Observing plants on a trip to Sicily, he had an arresting thought and imagined flowers as Proteus, the shape-shifter of Greek mythology: "While walking in the Public Gardens of Palermo, it came to me in a flash that in the organ of the plant which we are accustomed to call the leaf lies the true Proteus who can hide or reveal himself in all vegetal forms. From first to last, the plant is nothing but leaf" (quoted in Coen 2001, 524). He suggested that rather than being spaced along a stem, as are leaves, flowers are whorls—

concentric circles of plant organs, with leaves on the outer ring and the ovary at the center. He was right.

Again, as the splendor of the best bouquets makes clear, flowers vary greatly in shape, size, and color. But their unifying attribute is precisely those whorls of leaflike organs, highly modified for their particular purpose: pollination.

In this book, we tour the world in twenty-four hours, spending each hour with a different flower at a key moment in its day. Although you might not think it, there are flowers blooming and pollinators visiting them all through the day and night. As we travel together across time and to every continent, I hope you'll take a moment to reflect on and be amazed by the power and beauty of plants, the stories they tell, and what they teach us about resilience and survival.

Flower Day

MIDNIGHT
Moonflower
Ipomoea alba

(THE AMERICAS)

High in the trees, a pure-white flower rivals the moon itself. The moonflower, with its long tubular blooms, is a member of the group of plants called morning glories, but moonflowers are actually at their glory in the evening. They open at dusk and remain open all night, closing only when the sun comes up.

When open, the flowers are trumpet-shaped. Their flaring mouths look like white circles with five radiating green spokes.

Some compare them to flying saucers. They can be as large as a salad plate, with a long, narrow tube—sometimes as long as fifteen centimeters (about five inches)—protruding below and containing huge amounts of nectar at its base.

Their smell, reminiscent of suntan lotion or baby powder, is delicate, refined, and very alluring to the insects who visit the flowers in search of that cornucopia of nectar. As night wears on, moonflowers keep blooming, attracting hawk moths—large hovering moths also known as sphingids or sphinx moths. These heavy insects have the hover power of a hummingbird and long tongues that reach deep into the flowers to extract the highly nutritious nectar. Moonflower nectar has a very high sugar content, which hawk moths need to fuel their flying and hovering. At night these moths appear in abundance, swooping down on one another and muscling each other out of the way to win that sugary prize. But this is a transient feast. Once dew falls and dawn arrives, the flowers wither and collapse. They last only a single night.

We can find moonflowers in tropical areas all over the Americas, from Florida to Argentina. Humans cultivate them widely too; we love the mystery and fantasy of their flowers and, historically, the sap of their vines. Moonflower plants were a key component in the production of rubber in ancient Mesoamerica—their sap, when mixed with the latex of the *Castilla elastica* tree, created an elastic substance that could be formed, deformed, and made into objects like sandals and waterproof bags. The moonflower sap coagulated the latex, allowing Mesoamericans to make rubber hundreds of years before its supposed invention by European and North American scientists.

The Spanish conquistadores marveled at both the technology used to make rubber from tree latex and its products, such as bouncy rubber balls. The complex civilizations the Spanish encountered played highly competitive ball games that were an important cohesive element of Mesoamerican life, and some people in Mexico still played them well into the twentieth century. Even today, *huleros*, or rubber workers, in southern Mexico use latex from *Castilla* trees and sap from moonflower vines to make balls that bounce up to two meters (six feet) in the air!

In fact, rubber balls made with moonflower sap are sacred elements of the Popol Vuh, a K'iche' Mayan creation myth in which the Hero Twins, Hunahpú and Xbalanqué, played against the evil lords of the underworld—and won. They defeated death itself, then rose to the heavens to light Earth as the sun and the moon.

1 AM

Queen of the Night
Selenicereus grandiflorus

(THE CARIBBEAN,
CENTRAL AMERICA, MEXICO)

Refulgent CEREA!—at the dusky hour
She seeks with pensive step the mountain-bower,
Bright as the blush of rising morn, and warms
The dull cold eye of Midnight with her charms.
There to the skies she lifts her pencill'd brows,
Opes her fair lips, and breathes her virgin vows;
Eyes the white zenith; counts the suns, that roll
Their distant fires, and blaze around the Pole;
Or marks where Jove directs his glittering car

O'er Heaven's blue vault,—Herself a brighter star.
—There as soft Zephyrs sweep with pausing airs
Thy snowy neck, and part thy shadowy hairs,
Sweet Maid of Night! to Cynthia's sober beams
Glows thy warm cheek, polish'd bosom gleams.
In crowds around thee gaze the admiring swains,
And guard in silence the enchanted plains.
Drop the still tear, or breathe the impassion'd sigh,
And drink inebriate rapture from thine eye.

(ERASMUS DARWIN, *The Botanic Garden*, part 2, canto 4, lines 15–32)

Thus wrote Erasmus Darwin (grandfather of Charles) about the queen of the night in his extraordinary poem on the loves and lives of plants. It's hard to imagine, from looking at this cactus at most moments in its life, that it could inspire any poetry at all.

Found in dry forests and comprising grayish green stems that resemble a mass of snakes sliding over rocks and vegetation, the queen of the night seems like nothing special. Then a bud appears, growing over days to become a long tube topped by a bulbous mass of tightly packed tepals. The outer tepals—all the petallike structures of cacti are called tepals—of queen of the night flowers are pale tan or reddish brown and hide the opalescent-white inner tepals that are only revealed as the flowers open fully. The "admiring swains" of Darwin's rather torrid verse are the many pollen-producing anthers, and the "enchanted plains" refer to the many-parted pollen-hungry stigma—both also revealed when the flowers open up.

The queen of the night is a cactus but not, like the saguaro we will meet as the day closes, a cartoon image of one. In fact, the queen of the night was one of the first cacti to reach European greenhouses, and botanists cultivated it as early as 1700 in the

British royal palace at Hampton Court. The plant, and especially its flower, amazed all who stayed up late to see it bloom. When Linnaeus described this plant as *Cactus grandiflorus* in 1753, it had the largest flower of any cactus known at the time. Since then, botanists have found several other species of *Selenicereus* with larger flowers, but the queen of the night still holds a special place in the botanical pantheon as the first of these night-blooming beauties to be cultivated in Europe.

Let's watch as the drama of the queen of the night unfolds. As the sun goes down, the outer tepals peel back one or two at a time and the flower slowly opens—a sort of floral dance of the seven veils. Once open, the outer tepals bend backward, showcasing the inner white tepals, arranged in a cup that can be as large as forty centimeters (over a foot) in diameter—no wonder Darwin was impressed! By midnight the flowers are fully open and ready to receive visitors—and not just admiring botanists. The scent is reminiscent of vanilla or orange blossom, attracting moths and bats, who pollinate the flowers. The opalescent-white inner tepals may also reflect the moonlight and thereby attract other night-flying creatures searching for nectar at the bottom of the long floral tube.

By early morning the entire flower begins to droop and sag, and by sunrise all that remains is a flaccid remembrance of the queen of the night's amazing beauty. Each flower lasts only a single night, a fleeting reminder of the fragility of the natural world.

2 AM

Angraecum
Angraecum cadetii

(MADAGASCAR)

To understand orchids, think like a matchmaker. Their extraordinarily complex flowers have resulted in some of the most unusual relationships. Orchid flowers in many parts of the world have shapes and patterns that look like female wasps. Why? Sex sells. By tricking amorous male wasps into mating with them, orchids get the insects to spread their pollen.

There are even stranger plant partners. Consider the orchid *Angraecum sesquipedale*, which is found in Madagascar. The name *sesquipedale* is Latin for "one and a half feet" and alludes to the plant's extraordinarily long nectar spur. The existence of this spur led Charles Darwin to suggest that there must be a moth with an equally extraordinarily long tongue to drink that nectar. Sure enough, researchers found such a long-tongued moth and called it *Xanthopan morganii praedicta* in honor of Darwin's prediction. *Angraecum* species generally flower at night, have strong scents, and attract nocturnal hawk moth pollinators. Other hawk moths don't have tongues as long as *Xanthopan morganii praedicta*, but they're long enough. Imagine the orchid specialists' surprise when they discovered that some *Angraecum* species on the Mascarene Islands in the Indian Ocean were pollinated during the day—and by small songbirds!

Still more surprises were in store. *Angraecum cadetii* is an orchid found only in the wet forests of the Mascarene Islands of Réunion and Mauritius. It has a much less flamboyant flower than its relative from Madagascar. While the flowers of both species are creamy white, scented, and open all night, those of *Angraecum cadetii* are small and have tiny conical nectar spurs only about six millimeters long (about a quarter of an inch). Who pollinates this plant? Moths would seem to be the obvious culprits, but careful research using cameras and nightly observation revealed that the main visitor—and the principal pollinator of *Angraecum cadetii* on Réunion—was a raspy cricket operating under cover of darkness.

It's easy to assume that all insects or birds who visit flowers are pollinators. But many botanists argue that the title should be reserved for those who actually transfer pollen and fertilize ovules, helping to create the seeds of the next generation. Crickets usually eat plants, so this cricket's behavior—coming for the nectar, carrying away the pollen, and leaving the plant unchewed—seems unique. Careful experimentation has shown that these insects do indeed deserve to be labeled pollinators—when there are no crickets, there are no fruits, no seeds, and no new *Angraecum cadetii* plants.

Although the orchid *Angraecum cadetii* received its scientific name in 1988, the cricket remained anonymous until 2010. It was then dubbed *Glomeremus orchidophilus*, in honor of its role in pollinating the orchid. The insect is only found on Réunion, but the orchid is also found on Mauritius. So what gives? Well, as it turns out, there is another species of raspy cricket that lives only on Mauritius, and researchers are betting it will be a perfect match.

3 AM

East Coast Banksia

Banksia integrifolia

(AUSTRALIA)

In the dark night of the Australian bush, the creamy-white inflorescences of banksias glow against the trees' dusky leaves and stems. Who will pollinate these flowers, tucked away inside the branches of the plant? Surely not the local birds, who are not around this early in the morning. Neither are any insects up to the task. It turns out that the nectar and pollen from these flowers are a magnet for marsupials—specifically tiny pygmy possums and sugar gliders.

Banksias, along with wattles (*Acacia* spp.), are icons of the Australian flora. Found only on that continent, their numerous species are very diverse and grow in many vegetation types. Banksia flowers live in inflorescences—compact spherical or cylindrical arrangements of many small flowers that grow in concentric circles. East coast banksia inflorescences look a bit like floral feather dusters and can be ten to fifteen centimeters (almost half a foot) tall. Each flower is a minimalist structure, with only two pollen-producing stamens and a single pollen-receiving style. As the flower opens, the style elongates, eventually bursting through the petals to extend beyond the stamens; the protruding styles give the inflorescences their brushy appearance. The style is covered with pollen as it elongates, but the pollen-catching stigma hides away until later in flower development. The flower thus avoids self-fertilization except as a last resort. Flowering occurs relatively rapidly, with more than a hundred flowers opening per day, but continues over several days.

At the very base of the flower, the plant produces a sugary nectar, a resource for all sorts of animals, including humans. In Australia many different nations of Aboriginal people collect banksia inflorescences and soak them in water to make a sweet energy drink, which they sometimes ferment into a mildly alcoholic brew. Many other animals, including insects, birds, and mammals, visit the flowers—all also seeking its sugary reward. The flowers produce nectar all day and night, but nighttime is when tiny pygmy possums, with their prodigious climbing ability, clamber through the trees in search of it.

For a long time, scientists discounted nonflying mammals as plant pollinators; it was hard to imagine them reaching and then traveling between flowers. But studies of banksias in Australia—and of proteas, their relatives, in South Africa—have clearly shown these mammals' importance. The robust inflorescences of *Banksia integrifolia* can easily support the weight of the pygmy possums and sugar gliders, and because the cylinders are found at the junction of branches, often deep within the tree, they are accessible only to these nocturnal climbing creatures and not to the day-flying birds. As the possums and gliders clamber over the inflorescences, pollen dusts their bellies, moving with them from inflorescence to inflorescence over the course of the night. Although they're looking for a midnight nectar snack, the mammals—grooming themselves after or even during their forages—also find a rich resource in this pollen. It is high in nitrogen and an important part of their diet.

These nocturnal possums and gliders may not travel as far as birds like honeyeaters, who also visit banksias, but they are still important. Researchers in Australia have shown that these marsupials pollinate many of the flowers that birds fail to visit. This creates more seeds that will grow into more banksias. So while banksias are among the daytime glories of the Australian bush, their nighttime activities are also important for their success.

4 AM
Chicory
Cichorium intybus

(EUROPE, INTRODUCED TO NORTH AMERICA)

In the mid-eighteenth century, the Swedish botanist Carolus Linnaeus suggested that we use flowers to tell the time. Imagine planting his *horologium florae* (flower clock) as a garden, with flowers arranged into a clockface according to their opening times. He was working in Uppsala, Sweden, where the days begin very early in the summer. Midsummer sunrises occur at 3:30 AM! That flowers open and close at different times throughout the day was not a new idea; keen observers had long noticed this phenomenon. It is even described in this seventeenth-century poem extolling the beauty, quiet, and innocence of nature, as contrasted with the sinfulness of humans:

How well the skilful gard'ner drew
Of flow'rs and herbs this dial new,
Where from above the milder sun
Does through a fragrant zodiac run;
And as it works, th' industrious bee
Computes its time as well as we.
How could such sweet and wholesome hours
Be reckon'd but with herbs and flow'rs!
(ANDREW MARVELL, "The Garden," stanza 9)

The blue flowers of chicory, along with the yellow flowers of hawk's-beard and bristly oxtongue, were Linnaeus's choices for 4 AM. In Uppsala, early risers could see chicory's pale-blue blossoms—each composed of flowers themselves—just after sunrise. Linnaeus's three 4 AM species belong to the plant family Compositae, or Asteraceae, and their flowers are called composites because what looks like one flower is actually a group of tightly packed flowers. Daisies and sunflowers are classic examples of these "superflower" structures.

There are two types of flowers in composites: disk flowers, like those at the center of the sunflower we'll meet in a few hours, and ray flowers, which are strap-like and resemble conventional petals. Chicory has ray flowers, each of which looks like a blue strap with five little teeth at the tip. The chicory superflower has about twenty of these ray flowers, arranged in a circle such that each ray looks like a single petal. The dark-blue anthers and styles are clustered at the center, ready for insect visits.

While chicory flowers open early in the day, they don't start blooming until late in the growing season, after other flowers have packed up and set seed. Because of this, the chicory plant has become a symbol of persistence and waiting. It has also long been associated with magic, even credited with removing curses, granting invisibility, and unlocking doors. The plant is prized for its more practical uses too. Its roots, for example, are sometimes dried and ground into a coffee substitute. Chicory is a relative of lettuce (yes, even lettuce has flowers) and has many different cultivars—varieties that humans have cultivated for certain traits. These include the Belgian endive, also known as witloof chicory, whose tender leaves grow in the dark. The curly endive found in salads is a closely related but different species, *Cichorium endivia*. That goes to show how confusing common names can be!

Every night the entire inflorescence closes up, and no flowers are visible. As the day dawns, everything opens up again to welcome small hovering insects. Who specifically is visiting? Hoverflies, also called syrphids or flowerflies. These pollinators are incredibly important, thought to pollinate over 70 percent of European wildflowers. Some people confuse hoverflies, which live all around the world, with bees or wasps because many of them have yellow-and-black-striped bodies. Hoverflies are often called bee mimics, but maybe it's the bees that mimic them. If you want to know the difference, flies have two wings and big eyes that almost meet in the middle of their heads, while bees have four wings and smaller eyes. Syrphids also hover over flowers like tiny helicopters, probing for nectar with their long tongues or collecting pollen on their hairy faces to be groomed away and consumed later. Bees usually land on flowers to access nectar and pollen. Next time you want to swat at flies, thank them instead for their services in pollinating so many of the flowers in Linnaeus's *horologium florae*—especially if you're sipping some chicory coffee early in the morning.

5 AM
Dandelion
Taraxacum officinale

(WORLDWIDE)

Human beings have a love-hate relationship with dandelions. Harbingers of spring and the last flowers to leave us in autumn, they're also a bane for those who wish for a perfect green lawn. An entire industry has cropped up around getting rid of dandelions, producing tools such as the Dandelion Terminator—for me, a horrible thought. I love dandelions—bright and cheery, they remind me that a flower is only a weed where we might not want it. These spontaneous plants are ubiquitous even in cities,

making them ideal entry points for people to appreciate nature in all its myriad forms. They were here before we were and will be here long after we're gone.

The taxonomy of dandelions is fiendishly complex and highly contested. While the dandelions of lawns and roadsides are usually plants of the widespread species *Taraxacum officinale*, botanists recognize lots of other species too. In Britain botanists have named almost 250 dandelion species, while in North America they count only 15. Why the difference? It's not just due to location; it's also due to the peculiar reproductive behavior of dandelions and how their offspring are recognized. Dandelions can produce seed without pollination, giving rise to microspecies—populations that are genetically identical to and yet subtly different from their parents. Some botanists consider these microspecies distinct enough to warrant a scientific name, while others prefer not to distinguish these entities. But whether you're a "lumper" or a "splitter," not all dandelions are the same.

The English word *dandelion* comes from the French common name for this plant, *dent-de-lion* (lion's tooth, itself derived from the Latin *dens leonis*), referring to the plant's jagged leaves and their often backward-pointing lobes. Dandelion leaves are a prized salad ingredient in many cultures, both for their medicinal properties and as reminders of bygone times. As one of the seven ritual foods of the Passover seder, these bitter herbs or maror recall for those of the Jewish faith the bitterness of slavery and captivity in Egypt. The bitterness of the leaves comes from chemicals called sesquiterpenes, found in the milky sap throughout the plant. This sap has a diversity of uses, from getting rid of warts to mending bicycle tires!

From the rosette of leaves spring the dandelion flowers or, more correctly, inflorescences, for the dandelion is a composite; its flower heads are made up of many individual flowers packed closely together. Like chicory and lettuce, dandelions have ray florets with petals fused into a strap-like shape. Together, they look like a single fluffy bright-yellow disk. Individual florets open from the outer edge of the flower toward its center. The whole inflorescence closes up at night, opening again as the sun rises; on cloudy days it may not open at all. Because dandelions flower early in the season, they are very important sources of nectar and pollen for insects that emerge early in the spring. Weeders, take pity on the insects and leave the dandelions for their sake!

A dandelion inflorescence lasts only a few days; what follows is the familiar seed head. Each of the florets produces a single seed topped with a feathery parachute. Together, they form a fluffy white globe known as the dandelion clock. Why? Blow on the seed head and say "one o'clock." Are there seeds left? Blow again and say "two o'clock." Keep doing this until all the seeds are gone, and you'll know the time of day. At least that's what some English children used to claim. It's not particularly accurate, to be sure, but this tradition is the reason behind the dandelion's common name in England—one o'clocks. As a child I thought that if I made a wish and blew all the seeds off a dandelion head, my wish would come true. I can't say whether that worked, but even today, I never pass by a dandelion without blowing on its clock.

6 AM

Coyote Tobacco
Nicotiana attenuata

(NORTH AMERICA)

On an early summer's day just after dawn in the Great Basin of the western United States, flowers of the usually night-blooming coyote tobacco are opening, not wilting and fading as they would've had they been open all night long. Have these plants overslept?

Plants are sessile—they can't move and so live and reproduce in a single location. This means that they must develop strategies to avoid predation and attract pollinators. They do this with chemistry. Plants make an extraordinary variety of chemicals, many

of which humans have used as medicines, fragrances, and even food flavorings. But for plants, chemicals are essential for survival.

Coyote tobacco is what we call an annual plant—it lives for only one season. It grows in open spaces created after fires or other disturbances. Its tubular white flowers usually open at dusk, emitting a sweet scent that attracts their main pollinators, night-active hawk moths, who forage for the nectar contained at the bottom of the floral tubes. The main ingredient of this scent is the chemical benzyl acetone. Even if the plants have lots of sugary nectar, they need benzyl acetone to attract the hawk moths. But attracting hawk moths has a downside: the insects, after helping themselves to a sweet snack, lay their eggs on the coyote tobacco plant. These eggs hatch into voracious caterpillars that can strip a plant of its leaves, ruining its chances of surviving and therefore of reproducing. Thankfully, coyote tobacco has a defense plan.

When hawk moth caterpillar populations are large, consuming the leaves of many plants, coyote tobacco switches from flowering at night to flowering in the early morning, when there are fewer hawk moths around. How does the plant decide to make the switch? Chemistry is again the key. The caterpillars' munching prompts the plants to produce a chemical called jasmonic acid, which in turn causes their flowers to produce less benzyl acetone, have lower sugar concentrations in their nectar, and, crucially, open in the morning. Dawn brings hummingbird pollinators that do not lay eggs on the leaves. This increases the plants' chances of surviving and producing seeds for the next generation, but the hummingbirds are not as efficient as the hawk moths—they don't fly as far or visit as many flowers. A mixed strategy ultimately serves coyote tobacco best.

In addition to sugar, the nectar of coyote tobacco contains nicotine. Nicotine is found in all species of the genus *Nicotiana*, which includes *Nicotiana tabacum* (the plant used to produce cigarettes, cigars, and snuff). Experiments have shown that this chemical deters flower visitors from overstaying their welcome. Coyote tobacco flowers that were genetically manipulated to lack nicotine had more nectar removed from them, suggesting that pollinators spent more time on each flower. Nicotine, then, may induce animals to go from flower to flower more often, increasing the chances that they visit a different plant and enable genetic mixing. Nicotine also repels nectar robbers like carpenter bees, who drill into the bottom of the flower to extract the sugary reward but never encounter the anthers or stigmas to pollinate the flower. It's a subtle balance, with many trade-offs.

The chemical interaction of plant and animal doesn't stop at pollination. Too much hawk moth caterpillar chewing also stimulates the plants' production of chemicals that both upset the caterpillars' digestion and attract other animals, like birds or predatory wasps, that might eat the caterpillars! Coyote tobacco plants can identify the kind of caterpillar munching on their leaves by the saliva it leaves behind and then make chemicals that target the specific offenders. These plants are not just engaged in an arms race with their attackers; they are using chemistry to subtly change their behavior depending on who those attackers are.

Even though they can't move, plants aren't sitting around all day doing nothing. They are behaving—often in ways we simply cannot see.

7 AM
Curly Rock Rose
Cistus crispus

(AFRICA, EUROPE)

As the seasons change from winter to spring in southern Europe, the days grow longer, and plant life begins to speed up. For many years, I have been part of a field course along the Atlantic coast of Spain, in a place made famous by Miguel de Cervantes—Zahara de los Atunes, where some residents still use the traditional tuna-fishing method known as *almadraba*. This ancient technique involves placing nets underwater to capture the tuna as they migrate. My field course, too, gets caught every year during the blooming period of the rock roses, marveling at those iconic plants of the Mediterranean region.

Rock rose flowers last only a single day. When the sun rises over the Atlantic and hits the rocky promontories in Zahara, the rock rose flowers begin to open. It's a slow-motion spectacle. The fragile hot-pink petals of *Cistus crispus*—my personal favorite rock rose species—are crumpled inside the bud as if someone stuffed them in without regard to neatness or order. The first hint that a flower is about to open is a crack in the sepals that envelop the bud. The night before bloom, we see just a hint of pink through that crack. Shortly after sunrise, depending on the weather—things slow down if it's raining or cold—the five hot-pink petals begin to straighten out, gradually unfurling until they resemble a flat disk of iridescent, almost-fluorescent pink. The petals are darker toward the center, where a contrasting boss of yellow stamens is revealed. The wrinkles and creases slowly—but never completely—disappear. By late afternoon the petals begin to fall, one by one, carpeting the ground beneath the plants with pink.

Rock roses of different species often grow together and sometimes fertilize each other. If their flowers are open at the same time and the same pollinators visit them, the resultant seeds will grow into hybrids—plants that look like intermediates between their parents. *Cistus* species are very popular garden plants, and many of the varieties that we cultivate are hybrids of wild species. The crossing of species—hybridization—can sometimes produce intermediates that outcompete their parents, ecologically displacing or becoming more common than them, as happened with the salsifies we'll meet at noon. But hybrid rock roses in the wild don't produce fruit and so probably last only as long as an individual plant's lifespan.

On the slopes below the headland where *Cistus crispus* occurs in Zahara, there grows another pink species, *Cistus albidus*. The two plants are closely related and attract the same pollinators (mostly bees). They don't often grow near enough for the bees to easily carry pollen from one species to the other, but hybrid plants do develop—not many but enough to have kept my university field courses interesting for years.

In 2010 when the Eyjafjallajökull volcano in Iceland erupted, my field course was in Zahara. We were unable to leave, so a colleague and I mapped all the *Cistus crispus* and *Cistus albidus* plants and their hybrids in the area. The hybrids appeared only in a narrow area, where the two species co-occurred near headlands. More than a decade later, in 2023, in the very same area, I saw few of these intermediate plants—they seemed to have reached the end of their lives. A decade earlier, disturbances caused by the construction of roads and houses may have created just the right conditions for *Cistus albidus* populations to explode near the headland habitat of *Cistus crispus*. That density may have brought the two species close enough together to spawn the hybrid plants. When the habitat changed again, as building slowed from lack of space, opportunities to cross-pollinate became fewer.

Nature is always changing—even things, like our hybrid *Cistus* plants, that seem fixed. Hybrids can come and go as environmental conditions change or as human disturbances increase the chances that distant species meet. Our rock roses have changed with humans' expanded occupation and use of the Mediterranean habitat around Zahara. Rock roses were certainly there when Cervantes wrote about the *almadraba*, but their hybrids may not have been.

8 AM

Uape Jacana or Giant Waterlily

Victoria amazonica

(AMAZON BASIN, THE GUIANAS)

On the other side of the world, as the dawn chorus winds down and the sun comes up over the watery world of the Amazon, the huge flowers of the giant waterlily are slowly turning from white to pink and purple and closing over their centers. Trapped inside are the beetles who had come to the flowers during the night.

Just before sunset the night before, the prickly bud—about the size of a small can-

taloupe or a large mango—began to open, revealing many whorls of thick, white petals. Just a few days before that, this bud had been underwater, waiting to mature. Once flowering begins, the petals start bending backward to reveal the flower's center, which is surrounded by a complicated set of organs: There are several rows of staminodes—thin petallike structures thought to be derived from the stamens, the pollen-bearing organs, but with no pollen in them. Then there are the stamens themselves, each full of yellow pollen. Finally, atop the central whorl of ovule-bearing pistils, there are the carpellary appendages—fleshy structures full of carbohydrates and sugars. These are tasty treats for the beetles who, once night falls, are lured to the flowers by the heady scent they emit as they open. It smells like a mixture of pineapple and butterscotch, if you can imagine such a thing! The inner petals heat up, which helps spread the odor far and wide, attracting beetles from the surrounding forest. The beetles feast in the flowers all night and, failing to leave before sunrise, are trapped inside as the petals begin to close again.

In the morning, the giant waterlily's flowers are pale pink, darkening to purple toward the center, and partly closed. They no longer smell wonderful, but their starchy appendages keep the trapped beetles busy all day. Meanwhile, something else is happening inside the closed flowers: the stamens are releasing pollen. By the time the flowers open fully later in the afternoon, the beetles will be covered in it and—having eaten up the spongy carpellary appendages—ready to move on. The timing is perfect. As the beetles fly away, the next day's flowers are just beginning to open—temptingly white and lusciously scented on the water's surface. Because the female parts of the flowers are receptive on the first night, before the pollen is released, cross-pollination is guaranteed, increasing genetic diversity in and ensuring the future of the waterlily population.

Even when they aren't flowering, these plants are extraordinary. They are denizens of oxbow lakes along the Amazon, places where a meander of the river has been cut off to form a crescent-shaped body away from the current. The huge leaves of the giant waterlily cover the surface of the water, and their prickly undersides and upturned edges allow them to push other plants out of the way to capture as much sunlight as possible. Darlings of the tropical greenhouse, these waterlilies are more diverse than previously thought. The

COVID-19 lockdown drove botanists and botanical artists to spend more time examining the flower growing in one such greenhouse at the Royal Botanic Gardens, Kew, in London. They found that the plant in cultivation was a new species, *Victoria boliviana*, joining the two already known species from the northern Amazon (*Victoria amazonica*) and the Paraná (*Victoria cruziana*) river drainages. Even with a flower this big and spectacular, there is still much to discover beneath the surface.

9 AM
Sunflower
Helianthus annuus

(NORTH AMERICA)

A field of flowering sunflowers is an impressive sight. Like dandelions and chicory, sunflowers are composites: what we call a flower is really a group of tightly packed flowers. The "super" sunflower structure consists of both ray flowers, which have long strap-like petals, around its edges and disk flowers, which have five tiny petal lobes, at its center. Each disk flower produces a single seed. The huge bright-yellow flower heads not only look like our image of the sun but, when they all point east together, make us want

to watch its rise too. Is their movement just a collective response to available light, or is an internal clock—a circadian rhythm—involved?

Most organisms have some sort of circadian clock, a built-in response to a twenty-four-hour cycle. The rhythms of life are synchronized by it; it's what makes some people morning larks and others night owls. As we have seen, plants' circadian clocks are sensitized to and synchronized by—but sometimes independent of—light, the source of the energy they use to produce food to grow and ultimately flower. Put plants in the dark, and they'll continue to exhibit circadian behaviors, such as movement, germination, and fragrance emission—at least for a while. But plants also show direct responses to light. Many of us have put a houseplant in the window, only to find a few days later that all the leaves are pointing toward it, or done experiments in science class to germinate beans in the dark with a light source on one side. This type of movement—of plants' leaves, their light-capturing organs, toward incoming light—is called phototropism.

Some plants exhibit a particular form of phototropism called heliotropism, or solar tracking, in which they follow the sun's movement throughout the day. The scientific name for the sunflower, *Helianthus*, comes from the same Greek root. Helios was the Greek god and personification of the sun, who crossed the earth each day in a golden chariot pulled by white horses. So sunflowers must track the sun, right? The answer, it turns out, is both yes and no. As youngsters, sunflower plants are indeed heliotropic, pointing to the west at sunset and then reorienting themselves to the east overnight. A sped-up video would show them seemingly waving at us. But return to that same field at sunset once our flowers have matured, and they'll still be pointing east.

A complex interaction of a circadian rhythm and responses to light dictates this phenomenon. Even if kept in the dark, young sunflowers would still follow their usual movements for several days. To understand this, we need to look below the flowers' heads, at their stems. In the morning, growth rates are highest on the east side of the stems, making the plants bend to the west; at night, the rates are highest on the west side, causing the stems to face east again by morning. For sunflowers, when stem elongation stops, solar tracking stops too.

So why do mature sunflowers point east? This orientation maximizes morning-time flower-head warming and thus pollinator visits, leading to greater reproductive success. Pollinators are most active first thing in the morning, so it's not worth a full-grown plant's energy to track the sun to the west. Solar tracking, or the lack thereof, makes evolutionary sense at different points in the plant life cycle. Our mature sunflowers will still be pointing east this evening, but earlier in their lives, those growing young plants preferred a more varied view.

10 AM
Sago Palm
Metroxylon sagu

(SOUTHEAST ASIA)

By the time the tiny greenish-white flowers of the tall sago palm in the forests of the island of New Guinea begin to open, the plant is already past its prime for human use. The massive inflorescences of these palms can each be as big as a small car (think of an old-school Volkswagen Beetle) and are produced only once in a plant's lifetime. When an individual palm has flowered and fruited, it dies. Each inflorescence flowers over a period of about three months and can produce between a quarter of a million

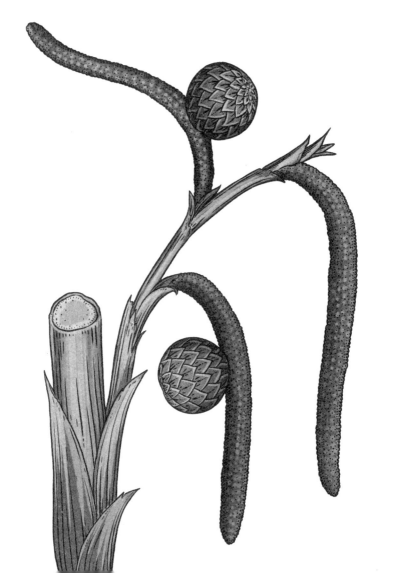

and a million fruits. Monocarpy—flowering and fruiting only once in a lifetime—is not that unusual; all annuals do it, as do familiar plants like agaves and some bamboos. The sago palm's flowering behavior, however, has been for millennia at the center of a vibrant culture that celebrates and respects nature.

Over the course of its life, the sago palm stores starch in its bole (trunk) to fuel its episode of flowering and fruiting. This starch is a staple for the people in the plant's native region, but sago culture is about more than just subsistence; palm groves are not only food sources but also sacred places. The starch is most abundant after a palm initiates an inflorescence but before it flowers. This is when harvesting begins. Sago palms have been planted in plantations, but true sago culture involves harvesting from the forest.

In Indonesian West Papua, sago harvesting is a communal event. Entire villages go to the forest to "know sago," singing traditional songs, listening to the forest, and becoming one with the environment and its multispecies interactions. Palms are carefully monitored, and when a palm is just ready, the village moves to the grove for the harvest. The palm is first selected, then it and the forest are asked for permission, without which the resultant starch will not be of high quality. The palm is next felled as close to the base as possible, using a traditional adze, a tool similar to an axe. When the palm falls with a resounding thump, the harvesters eat the cabbage, or heart of palm, before the real work begins. Taking turns with machetes and adzes, they split the bole and remove the starchy pith. They then rinse the pith and filter the water using the base of one of the palm's own leaves. The starch settles in a container, also made of sago leaves. They use the whole plant in the harvest, which can last for a few days to several months. There is no waste here.

Despite only living to flower once, a sago palm produces suckers from its base. When cut away from the main stem, these pups can be regrown and transplanted to make new trees. Although the Marind people of West Papua care for and cherish the palms and palm groves, they don't see their work as cultivation. Sago is a wild plant, part of the forest and its interconnectedness. The starch it produces is emblematic of the social relations between people and nature across time, sometimes through the stories of past harvests that are shared while in the forest.

But the inexorable spread of oil palm plantations threatens this sustainable way of life and sago culture itself. As monocultures—farming that focuses on growing only one type of plant in an area—these plantations lack the interconnectedness that is characteristic of sago groves, and they disrupt the storied way in which people's lives are attuned to the forest.

Oil palm killed the sago
Oil palm killed our kin
Oil palm choked our rivers
Oil palm bled our land.

(song of MARCUS GEBZE, Elder of Mirav village, West Papua; quoted in Chao 2022, ix)

Considering the agency of plants and all life is central to sago culture. The oil palm, despite its shortcomings, still has agency in the eyes of the people of West Papua. Compassion and pity sit alongside resentment and fear; this alien species, with no relatives and no connectedness, is a victim too.

11AM

Western Blue Flax or Lewis Flax

Linum lewisii

(NORTH AMERICA)

Midmorning in the Rocky Mountains in western North America reveals slopes of pale blue, shimmering in the light. This wash of color is from the delicate plants of the native western blue flax. Blink and you'll miss the show: the petals of these plants will all have fallen by early afternoon.

Early explorers of the American West

consistently reported seeing fields of flax, and Meriwether Lewis, whose expedition across the Louisiana Purchase in the early 1800s documented the westward extent of the United States, wrote of the plant's potential in his journals. He had seen many Native peoples use it to make snowshoes, baskets, and fishing nets. The stringy fibrous stems of western blue flax, like those of the flax cultivated in Europe, are perfect for making cordage and string. Common flax (*Linum usitatissimum*), an annual plant, is the source of linen. Lewis noticed that the flax he saw was a perennial, sprouting year after year from a woody base, and thought that it might be a better crop than the annual species.

The flowers of western blue flax look like flat plates with five pale-blue petals, which often have darker stripes toward their base. The buds droop downward, straightening up only in the evening before flowering. Each petal grows from the base beneath the ovary and stamens. As the flowers begin to open, the pollen-producing anthers stand straight, directly above the five threadlike pollen-receiving styles at the center. But as the day continues, the anthers arch outward. These delicate flowers produce only a tiny amount of nectar as a reward for—and no scent to attract—potential pollinators, but they are nonetheless visited by a wide variety of insects.

Botanists who have followed in Lewis's footsteps and studied these flowers have found that small bees and flies are their most common visitors and their most effective pollinators. Western blue flax flowers are self-fertile, meaning that they can produce seeds even when ovules receive pollen from the same flower. Still, they need insects to move that pollen around—from anthers to stigma, and preferably from one plant to another to promote genetic diversity.

Many insects—including common fly pollinators, like hoverflies and bee flies—rarely travel to high-arctic or high-elevation systems. But in the Rocky Mountains, muscoid flies dare to go where few others have gone before. These forgotten relatives of the common house fly carry as much pollen on their bodies as do the more common fly pollinators and can be critically important when other insects are scarce. These flies dominate western blue flax pollination at such high elevations.

Unlike the intricate orchids we saw at 2 AM, which require very specific pollinators, the western blue flax can attract a wide variety of insects thanks to its wide open and seemingly plain structure. This is perhaps the reason for its success across a wide geographical range. The plant can take advantage of whatever insect community is present wherever it grows: bumblebees at middle elevations, smaller bees and flies at higher elevations, and muscoid flies at the highest elevations.

No matter where the plant grows, though, its flowers lose their petals by midafternoon. One by one the petals fall, whether or not the flowers were successfully pollinated. If they were, their seed pods will produce hundreds of tiny oil-rich seeds, holding the promise of future flowers for future blue mornings to delight future flies.

NOON
Jack-Go-to-Bed-at-Noon or Meadow Salsify
Tragopogon pratensis

(EUROPE, INTRODUCED TO NORTH AMERICA)

Common names for living things can make identification difficult. They differ between languages and regions, and sometimes the same name is used for completely different species. But occasionally they are uncannily descriptive and tell us much about the plant in question. The common name in England

for the meadow salsify—Jack-go-to-bed-at-noon—is one such elucidating epithet.

Meadow salsify, like the dandelion, chicory, and sunflower we met earlier in the day, is a composite—what looks like a single flower is in fact an inflorescence of densely packed pale-yellow ray flowers. The central flowers have blackish-red anthers and styles. The dandelion-like flowers of salsify open in the early morning, and by noon the inflorescence is tightly shut with not a single flower visible, hence the common name. Once the inflorescence is closed, each of its flowers develops into a single seed, topped with a tuft of hairs. The fluffy seed head looks like a parachute, giant dandelion, or—to appropriate one of the plant's other common names—goat's beard. Look closely and you can see that the radiating hairs of the parachute are joined together by finer side hairs, creating a very efficient device for floating away on the wind.

Tragopogon species are native to Europe, but humans have brought them into North America, probably accidentally as seeds contaminating the grain stores of European immigrants. In the early twentieth century in the Palouse region of Washington state, botanists found strange salsify plants that looked like a cross between two common, weedy introduced species: our *Tragopogon pratensis* and its relative *Tragopogon dubius* (yellow salsify). Subsequent genetic study showed that these new plants were in fact hybrids. Botanists named the new species *Tragopogon miscellus*. The crossing that led to *Tragopogon miscellus* appears to have happened many times in many places, most often with the yellow salsify as the pollen donor (father) and the meadow salsify as the pollen receiver (mother) but sometimes the other way round. When the hybrid was first described, its populations were said to be "small and precarious," but *Tragopogon micellus* is spreading; it is now a major weed where it occurs. Quite unlike the rock rose hybrids we saw in the morning, salsify hybrids can reproduce by seeds and spread far and wide. But more detailed genetic analysis suggests that multiple introductions from different areas of Europe have contributed to the multiple origins of this new species. New ideas emerge as we delve deeper into plant genomes, allowing us to learn more and more about how evolution is happening day to day.

Meanwhile back in Europe, the decline in traditional hay meadows is confining meadow salsify to disturbed areas—more like what we would call a weed. We often see weeds as a problem, but they are overlooked players in the world of plants. Weeds can contribute to the creation of diversity and are often the result of our land-use practices. They are intricately connected to us.

1 PM

Antarctic Hair Grass

Deschampsia antarctica

(ANTARCTICA, PATAGONIA)

It's Midsummer Day on the snowy islands off the western side of the Antarctic Peninsula, and the sun is just beginning to dip from its zenith at forty-three degrees above the horizon. At this time of year, the day is long, with more than twenty hours of sunlight, but the temperature fluctuates wildly between sunlight and shade. Here, at the southern tip of the earth, plant diversity is

extremely low. But plants do still grow and flower on these islands.

One such plant is the Antarctic hair grass, which forms tiny, tufted mats about two centimeters (almost an inch) high that peek out from between rocks and gravel in the open places where snow and ice are absent. This little plant holds the world record for southernmost distribution—these grasses have been found growing on Alamode Island off Graham Land at latitude 68°43' S! Though the days are long in the Antarctic summer and the sun only just touches the horizon at midnight, the growing season is short. Plants must flower and reproduce rapidly in very difficult dry, windy conditions.

Grasses have flowers too. They are tiny but perfectly formed. Antarctic hair grass flowers occur in small inflorescences, composed of only a few individual flowers, which are called florets in grasses. The florets themselves occur in spikelets. (Plant families with such specialized and distinct structures usually have their own terminology.) The flower itself—simplicity personified—has just two or three stamens with dangling pollen-producing anthers and one feathery stigma that protrudes to catch windblown pollen with great efficiency. At the base of the stamens and ovary are three minuscule bump-like projections, the equivalent of petals in other flowers. Each floret produces a single seed and only under long-day conditions, so it is important for these plants to flower as early in the season as possible.

The fierce wind of the Antarctic islands surely helps move hair grass pollen around, but these plants do not rely solely on sexual reproduction from seed. They also spread vegetatively via tillers, branches that produce their own roots to become independent plants. Hair grass is one of the preferred materials, along with mosses, for gull nest construction and is carried long distances by the birds. Areas around gull nests are high in nutrients—perfect for establishing new grass populations!

Antarctic hair grass is also found in South America, in Patagonia, where the grass populations are more genetically diverse than those on the islands. Researchers have suggested that the grass may have originally reached Antarctica by seed dispersal in warmer times or transport of stems via birds, or both. And the spread of this grass continues. Tillering increases with warming temperatures, and the Antarctic Peninsula and its western islands are warming faster than the continent itself; in fact, this region is thought to be the fastest warming area on Earth. Monitoring over the latter half of the twentieth century showed that mean summer temperatures increased by about one degree centigrade and the number of individual Antarctic hair grasses increased up to two-hundred-fold! Although global warming might seem to be a good news story for the Antarctic hair grass, that is far from the case. Rising temperatures caused by human-driven climate change can also interfere with the plants' fine adaptations to their extreme environment, disrupting their ability to survive long enough to reproduce in the first place.

2 PM
Titan Arum or Corpse Flower
Amorphophallus titanum

(SUMATRA)

Every time a titan arum blooms in a botanic garden, it is a sensation. People queue around the block to see this gargantuan vegetable wonder and to experience the smell its blooming produces. The other common name of this plant, corpse flower, hints at the nature of that scent. Each flowering event lasts but a couple of days, and the staff at botanic gardens are becoming experts at predicting when a plant will bloom. This

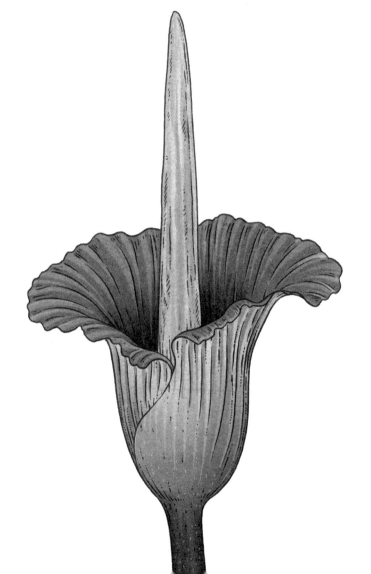

helps them prepare for the spectacle, focus visits, and avoid public and media disappointment. But why so much furor over the flowering of this plant?

The titan arum is truly a marvel. When the plants are not flowering, they produce a single leaf whose stalk can be over twice as tall as the average adult human and as thick as a small tree trunk, topped by a deeply divided umbrellalike blade under which a family of four could easily shelter from the rain. These plants grow from underground tubers and never produce leaves and flowers at the same time. They store their energy as starch in the tuber to fuel flowering events. The bloom is, like many of the flowers we have met so far today, technically an inflorescence—but this time the group of flowers is enclosed in a leafy structure known as a spathe. The titan arum spathe is pleated and ruffly at its edges and embraces the columnar spadix; at its base there are concentric rings of hundreds of tiny, minimalistic flowers with no petals or sepals, only functional reproductive parts. First are rings of flowers with a bright reddish-orange flask-shaped pistil and a purple, green, or red style and stigma. Above those are rings of pollen-bearing flowers, each with a single stamen. Then finally, at the top of the spadix and emerging beyond the spathe is a large, fleshy yellowish-green structure that can be several meters tall. This is the part of the inflorescence that a visitor sees—and smells. This yellowish-green column emits a cocktail of sulfide compounds over the course of flowering, producing an odor like that of rotting animal flesh, rotten eggs, or pickled radishes and ultimately culminating in the stench of excrement. The column heats up to about 38°C during the early flowering phase, helping spread these volatile compounds far and wide. Phew.

These smells not only cause shivers of delighted disgust to run down the spines of botanic garden visitors; they have a purpose in the rainforests of Sumatra, where the titan arum grows naturally. There, deep in the forest understory, the assortment of odors emitted by the fleshy spadix almost certainly serves as an attractant for insects, whose services are needed for pollination. Observations of titan arum blooms in the rainforest are few and far between, and researchers have yet to perform detailed studies of pollination in the wild; in fact, it was once thought quite wrongly that elephants pollinated the plants! But botanists have found many different small insects—beetles, fruit flies, bluebottle flies, and stingless bees of the genus *Trigona*—covered in pollen among the flowers at the base of titan arum spathes in Sumatra. Whether these insects are actual pollinators or just visitors is not known, but it seems reasonable to suppose that at least some of them transfer pollen from staminate (male) to pistillate (female) flowers.

Flowering occurs over two days. On the first day, in the afternoon, the spathe opens up to its full extent and is turned back like a ruffly maroon-and-green collar. At night the pistillate flowers bloom. It's at this time that the stench of carrion is at its strongest. In the morning the spathe closes a bit, then opens again in the afternoon—ready for the staminate flowers that bloom during the second night to release their copious amounts of pollen. The staminate flowers also generate heat, and researchers have suggested that the associated infrared radiation attracts insects. Once the staminate flowering phase is over, the spathe closes again around the developing fruits. The fruits mature inside the spathe until it rots away to reveal what looks like a column of tiny tomatoes.

As the spathe disappears, so too does the queue of human visitors. But like that of Proust's madeleine cakes, the unforgettable smell of our corpse flowers ensures that the memory remains.

3PM

The Traveler's Tree
Ravenala madagascariensis

(MADAGASCAR)

In the lazy midafternoon hours, the forests of Madagascar are full of activity—some of which, as you might expect on an island of fabulous diversity, is quite unexpected. High in the canopy of the coastal forests, a lemur leaps from a branch and lands on a most unusual plant with huge banana-like leaves arranged in a perfect fan. The lemur is looking for the flowers of *Ravenala madagascariensis* and their rich nectar.

The traveler's tree is the national tree of Madagascar and takes its scientific name from the Malagasy word *ravinala*, meaning "leaf of the forest." These extraordinary plants—extraordinary even in a land of botanical wonders—are not truly trees, even though they can reach heights of thirty meters (some one hundred feet). As anyone who counts the rings of a tree to determine its age can attest, trees grow wood by expanding the width of their stems and branches. But *Ravenala madagascariensis* is a monocot, related to lilies and grasses, and monocots do not have this secondary outward growth. So the traveler's tree is really a giant herb! It resembles its closest relatives, gingers and birds-of-paradise. Between its leaves, which fan out in a flat plane, arise stout, chunky inflorescences. Each inflorescence is an overlapping set of tough, boat-shaped green leaves, called bracts, that hold a few pale, spiky flowers.

Our black-and-white ruffed lemur clambers up the fan of leaves using their overlapping bases as a ladder and approaches the inflorescence from above. The bracts of *Ravenala madagascariensis* are thick and strong enough to support the animal's body weight of about four kilograms (eight pounds). Peering into a bract, the lemur finds a flower and, balancing on its hind feet, uses its front feet to pry the petals open. *Ravenala* flowers do not open when they bloom; instead, by keeping their petals closed at the tip, they hold their anthers in tension. As our lemur pulls the flower apart to access the rich nectar at its bottom, the anthers spring forward with a pop, dusting the lemur's head and muzzle with pollen. When our lemur visits its next flower, the pollen will be brushed from its head onto the flower's stigma, then it will grow down the flower's style and fertilize its ovules.

The flowers of *Ravenala madagascariensis* are creamy white and produce nectar day and night. For this reason and for a long time, botanists assumed that bats pollinated them, and it is true that bats often visit cultivated traveler's trees. But in eastern Madagascar, we don't see bats visiting the flowers. And the sunbirds who do visit are just visiting—they don't touch the stigmas or anthers. Black-and-white ruffed lemurs, however, visit every flower on a plant and often move between inflorescences.

There is still a lot left to find out. Although the Malagasy give many different names to *Ravenala* across the island, botanists had long thought that there was but a single species. Recent work, however, has described five new species, all corresponding to the varieties recognized by the locals. These huge plants are difficult to collect, and the geographic ranges of each new species are not well-known. They all have at least one thing in common though: lemurs. Black lemurs visit and lap up nectar from *Ravenala agatheae* in western Madagascar, covering their heads in pollen just as our black-and-white ruffed lemurs do with *Ravenala madagascariensis* in the east. But all is not well. Both lemur species are at risk of extinction, and thus this relationship between plant and primate is endangered too.

4PM
Four-O'Clock
Mirabilis jalapa

(THE AMERICAS)

I can easily imagine the hot-pink flowers of the four-o'clock as perfect attractants for birds or bees—or perhaps hummingbirds, who often visit bright pink or red flowers in the Americas. But, as we've seen, nature is full of surprises. These flowers open late in the afternoon, hence the common name, and remain open all night, attracting moths— mostly long-tongued hawk moths, who can reach the nectar that's held deep within the flower—instead. Not all night-blooming flowers are white!

The brilliantly colored flowers of the four-o'clock are unusual in other ways as well. What looks like petals is actually a calyx, the outermost whorl of a flower that is usually green and leaflike. We may also see flowers of different colors and variegation patterns on the same plant—all in vivid shades of yellow, hot pink, or red. The flowers often change color, from yellow to magenta or from white to pale pink. This metamorphosis, combined with the plant's heady fragrance at dusk, has made *Mirabilis jalapa* a garden favorite. But four-o'clocks have historically been grown for more than their sensuous beauty; the Aztecs in Mexico used them as medicine. People take plants to many places for many reasons, so their native distribution can be difficult to untangle, but as the four-o'clock's other common name—marvel of Peru—suggests, the plant is native to the dry forests of South America.

Four-o'clocks—along with their relatives, such as carnations, cacti, bougainvilleas, and beets—owe their bright pinks and fluorescent yellows to pigments called betalains, which are found in only these, and no other, flowering plants. Four-o'clock flowers also come in variegated forms: yellow with pink dots or stripes, hot pink with paler pink stripes, and all patterns in between. This variegation is apparent in wavelengths both visible and invisible to us, increasing the attractiveness of the flowers to pollinators who can perceive a greater range of light—the visual equivalent of a dog responding to a high-frequency whistle.

Scientists are attracted to the flowers' variegation for something else invisible to the gardener—their genes. Genetics govern variegation, and the four-o'clock flower's genome is itself like an ecosystem, full of phenomena scientists are only starting to unravel. *Mirabilis* has contributed to genetics research in another way too. Gregor Mendel's famous experiments with peas demonstrated the mechanism of inheritance, but his lesser-known work used four-o'clocks to show conclusively that a single pollen grain can pollinate only a single ovule. Because the pollen grains are large, and the plant's ovaries contain but a single ovule, he was able to hand-pollinate four-o'clock flowers knowing exactly who the parents were. The first- and second-generation offspring's flower color ratios confirmed his idea that single gametes (pollen or ovules, like sperm or eggs) were the elements that conferred traits. He never published these experiments, but he described the work in his letters to the Swiss botanist Carl Wilhelm von Nägeli. Nägeli was not a fan of Mendel's ideas, however, and is thought to have discouraged him from continuing his work on inheritance. Not very collegial. Maybe he was jealous.

Advances in science are often seen as individual efforts, the results of a single great mind. This is not the case; they require collaboration and cooperation. As with pollination, research needs interactions before seeds form and new things grow.

5 PM
Harakeke or New Zealand Flax
Phormium tenax

(AOTEAROA—NEW ZEALAND)

When is flax not flax? When it comes from a totally different world, of course. When Europeans first encountered the people of Aotearoa, they marveled at the fine textiles that they used for capes and clothing. So, with the hubris of the colonizer, they called the plant that provided the fibers "New Zealand flax"—but harakeke, to give the plant its proper Māori name, has nothing to do with flax botanically.

Like the traveler's tree we saw earlier today in Madagascar, harakeke is a monocot—a rosette-forming plant with long strap-like leaves and a tall, imposing inflorescence of waxy yellowish flowers that sometimes look a bit like they're upside down. European flax, like the Western blue flax we saw just before noon, has small leaves and blue flowers. It has nothing in common with New Zealand flax beyond both plants' role as the source of fibers in a local culture. In Aotearoa people brought harakeke along wherever they went; its utility for making cloth made its value second only to that of plants used as food.

Preparing the leaves to extract the fibers is laborious. Harakeke is grown in plantations called *pā harakeke* and is always approached with respect before harvesting. Only the outer leaves—likened to members of the older generation—are cut; the inner leaves, called *rito* or children, are left to be used another day. Once the thick leathery leaves are cut, they are scraped thoroughly with a mussel shell to separate the leaf tissue from the fine white fibers. Rolling these fibers on the thigh softens them before they are boiled, dried, and sometimes dyed. It may look easy, but it's hard and time-consuming work. The result—beautiful, soft yet strong cloaks with striking black-and-white patterns woven into them—so impressed early European sailors like Captain James Cook that they took the plants back to England in the hope of using their fibers to make sails, which were so crucial in the days before steamships. Convicts sent from England to the brutal penal colony on

Norfolk Island (today part of Australia) in the late eighteenth century were tasked with creating a flax industry from *Phormium*, but not a single person skilled in the preparation of the fibers was among them. The British kidnapped two Māori youngsters in an attempt to import the technology, but they were young men and it was women who knew how to prepare the flax. The entire enterprise was a dismal failure.

The flowering spikes of harakeke are stout and tough. Once the spikes begin to grow, they rise rapidly and don't stop until they're high above the fan-shaped clusters of leaves. One year the inflorescence on the plant in my front garden grew from what seemed like nothing to more than twice my height in less than a week! With its delicate side branches of red and waxy tubular flowers of yellow, the blooming plant is a spectacular sight. In harakeke's native land, its main pollinators are the tūī and korimako birds—honeyeaters found only in Aotearoa. Both these birds come to the flowers in search of the copious nectar at the bottom of the tubes, and their brushlike tongues are perfect for lapping it up. Perching on the inflorescences' thick stalks, they bend their bodies at just the right angle to reach deep into the curved flowers and gain access to their reward. As this is happening, the anthers, off to one side of the flower, dust the birds' foreheads with pollen—now ready to be carried to the next plant. The beaks of these birds are slightly curved to match the curvature of the flowers, which provides protection from nectar robbers who might not hit the anthers just right to then bear pollen from one flower to another.

Harakeke flowers are protandrous:

When the flowers first open, the anthers protrude from the floral tube, their bright-orange pollen contrasting strongly with the yellowish-red petals. As the day wears on, the style elongates until it reaches beyond the anthers, curving outward so that it's the first thing a bird's pollen-covered forehead touches. This improves the plant's chances of cross-fertilization and achieving genetic diversity. But the birds visit flower after flower on the same inflorescence, meaning that self-pollination is still probably the norm. Today the tūī and korimako have to compete for nectar with introduced starlings and mynahs—but these particular invaders are not the main threat. Ever since European colonists set foot on these lands, endemic species of all kinds have been on the decline in Aotearoa.

6 PM
California Poppy
Eschscholzia californica

(NORTH AMERICA)

It's mid-March, and the sun is setting over fields of gold in Southern California. The delicate-looking petals of the California poppies that cover the hillsides during the short but dramatic flowering season belie the toughness of these iconic flowers. By sunset the petals have all closed over the anthers at the center of the flowers, ready to open again the next morning and welcome the

bees who nourish their growing broods with the flowers' pollen.

The California poppy has been the state flower of California since 1903, and no other wildflower so epitomizes the phenomenon that is the "superbloom"—a mass flowering that cloaks the hillsides in color. The flowering season for poppies is short—only a few weeks—and hotly anticipated, especially in years when rainy winters seem to foretell extra special events. Locals and tourists flock to parks and reserves to witness the golden hillsides, so much so that in the spring of 2019, the area around Walker Canyon near Lake Elsinore in Riverside County was so crowded that it was a safety risk for people and wildflowers alike. All access roads and parking lots were closed. They were closed again for the 2023 flowering season, in anticipation of another massive event. And an event these blooms are—people are advised to track them on social media, and those who attend should take care to leave no trace behind. The poppies are flowering for a reason: to reproduce themselves. While such mass flowerings may seem to indicate that populations are in robust health and that nothing can harm them, climate change is affecting California poppies in different parts of their wide-ranging native habitats.

When spring seasons are warmer than average, the flowers bloom earlier, independent of rainfall. Populations from warmer, drier regions already flower pretty early and may not be able to flower any earlier—the plants would be very small, and small plants don't have enough energy stored for fruit production. But populations from cooler areas respond more dramatically to such increases in temperature. As the climate warms, the range of California poppies is predicted to shift toward these cooler regions at higher latitudes and higher elevations. The coordination of flowering and pollinator availability will then be more critical—and more precarious—than ever. If the timing is off, pollinators might miss the short blooming season and fruits might fail to develop, thus limiting the seed that will grow into new poppies to flower in subsequent years.

The flowers of California poppies can be bright orange or yellow, or any shade in between. They vary locally, but the most common color on the hillsides is orange. The opening of the poppy flowers is a sight to behold—the two sepals are fused into a structure that looks a bit like a pointed hat, and as the four petals expand, they push the hat off, freeing themselves to wave gently in the wind. The petals are very pliable and highly reflective at certain angles; they look like iridescent orange silk. This silkiness is produced not by the color but by the shape of the cells on the petals' surface. The cells of the petals' upper surface are elongated and have triangular cross sections, creating a series of ridges that act as prisms to focus light. The resultant silky effect, set starkly against the dull background of other vegetation, may be attractive not only to us but also to pollinators.

At day's end, the silky petals fold in on each other, creating an orange cone that's ready to open again at sunrise. This opening and closing takes place over the course of several days and explains the common name used by early Spanish settlers in California—*dormidera*, "the sleepy one." A superbloom of poppies is a sight you don't want to miss, but timing is everything. Our changing climate may make these events things of the past.

7 PM
Night-Blooming Jessamine
Cestrum nocturnum

(THE CARIBBEAN)

Like clockwork, as night falls, the flowers of the night-blooming jessamine open and begin to emit what some call the most powerful scent produced by any plant. The fabulous odor is reminiscent of jasmine—a combination of sweet and fruity, with what some perfumers call a note of dead-sexy muskiness. We can smell it from a considerable distance.

It's the kind of scent you either love or

hate; some people react to it with nausea or dizziness, while others drink it in with pleasure. This powerful nighttime odor has inspired the plant's many common names from all over the world. It is known in India as *raat rani* or "queen of the night" and in its native Caribbean as, among other names, *galán de noche* or "gallant of the night." Jessamine flowers open only at night, and many botanists long thought that it was the onset of darkness that stimulated their opening and the release of their odor bouquet. But, as with the sunflowers we saw earlier today, it turns out that a circadian rhythm governs flowering and odor production. Keep the plants in the light all the time and they will still bloom and share their scent according to a roughly twenty-four-hour cycle.

    The whitish-green flowers have a long, thin tube and five triangular petal lobes, which are perpendicular to the tube at flowering; they look a bit like miniature Olympic torches. Like most tubular flowers, these produce a sugary nectar from a structure at the base of the tube—and it takes an animal with a long tongue to reach that reward. The chemical cocktail that makes up the airborne odor contains a number of compounds that vaporize easily: benzyl alcohol, for example, suggests jasmine; and eugenol smells of cloves. This heady perfumed mixture emanates from the tips of the petals—cut them off, and the flowers are scentless; float the cut tips in water, and the scent continues to be released.

The odor of jessamine flowers attracts moths, who presumably pollinate the plant as they probe deep into the tubes to reach the nectar. The scent is strongest during the first couple of hours after the flower opens, a period that coincides with the peak of many moth species' activity. Its powerful nature—and its consequent detectability from afar—has inspired some farmers to use the flowers to lure moths away from economically important crops. The caterpillars of cabbage looper moths can eat three times their body weight in day, causing huge damage to commercial crops like mustards and cabbages. Trapping these pests before they lay their eggs reduces the damage they and their larvae can cause. And using a flowering plant to combat an insect pest is surely a better option than spraying toxic pesticides.

A member of the nightshade family—many of whose species, like tobacco and henbane, are toxic for humans—night-blooming jessamine also has a reputation for being poisonous. Eating the spongy white fruits can cause severe stomach pains, and some people are nauseated by the intense floral scent. The plant can be toxic to more than just human beings. Extracts of the leaves contain high concentrations of saponins, soapy bitter-tasting chemicals toxic to mosquito larvae. Because of this, humans often cultivate night-blooming jessamine in their backyards as a locally produced, plant-based mosquito control agent. Some have even considered using the plant to control malaria by attacking the mosquitoes that spread the disease.

But night-blooming jessamine itself is a bit of a pest. People have taken it all over the world, cultivating it for its scent. It's easy to grow and doesn't even need those moths to pollinate its flowers—in fact, jessamine doesn't need pollination to create fruit. Its white Styrofoam-like berries are eaten by birds, who disperse the seeds far and wide.

The seeds can sit for a long time in the soil, just waiting for space and sunlight. So when the plants escape from tropical gardens or greenhouses, they can spread rapidly and outcompete native shrubs and herbs. On the islands of the Pacific, including Hawaii, jessamine is classified as a high-risk weed that can form impenetrable thickets that choke out native vegetation and prevent the growth of rare and endangered trees.

A perfume to some, a poison to others, both a protector and a competitor—night-blooming jessamine, with its good and bad sides, is a Janus-faced plant.

8 PM
White Campion
Silene latifolia

(EURASIA, INTRODUCED TO NORTH AMERICA)

As dusk falls in the northern hemisphere, the white campion's flowers begin to open, revealing their bright white petals. But wait, although these sticky herbaceous plants look the same, there seem to be two different types of flowers—each on a different plant. One has five long, thin styles protruding from the ovary out the top, and the other doesn't; instead, it has ten pollen-bearing anthers hidden inside.

Most flowers have both male and female

sex organs in the same structure and produce both pollen and ovules, which go on to become seeds. White campions, however, are among the 6 percent of flowering plants that are dioecious, meaning that each individual is either female (pistillate) or male (staminate). This sexual system, while relatively rare, has evolved many times. White campions are unusual in that they have sex chromosomes like animals; in fact, white campion chromosomes are remarkably like those of humans. Female plants who go on to produce seeds are XX, and male plants who produce only pollen are XY; their Y chromosome shows gene loss and degeneration, just as in the Y chromosome of humans. Because the sexes are on separate plants, white campions rely completely on pollinator visits for fertilization—they cannot fall back on self-fertilization as many other plants, including the night-blooming jessamine we just met, can.

White campions are native to Europe, but humans brought them to North America more than two hundred years ago. Their most important pollinators are moths that probe the flowers for nectar as dusk falls. Moths visit male and female flowers randomly, ensuring that pollen is deposited on the stigmas. In North America moths are efficient and effective pollinators, but in Europe their service comes at a cost.

In the white campion's native range, pollination falls to the small, brown lychnis moth. In a bizarre twist, the female moths lay their eggs on the flowers' ovaries. The developing caterpillars then eat the seeds, severely limiting the reproductive potential of any plant unlucky enough to have eggs laid in its flowers. In fact, the caterpillars destroy a significant number of white campion fruits, leaving only some of the flowers to survive and produce seeds. Moths don't lay eggs in the staminate flowers but do still visit them and carry their pollen from flower to flower. White campions attract lychnis moths with their fragrance; the odor advertises their flowers as both sources of nectar and prime sites to lay eggs. This nursery pollination system clearly has a cost, but male moths also visit the flowers to sip at nectar without posing a threat to the plant. There is even evidence that the plants can change their scent after pollination to deter further egg layers from visiting. As we surely know by now, the economy of nature can be quite baffling, and the interactions between flowers and their visitors are rarely simple and straightforward.

Populations of white campion in North America have escaped this interaction and become invasive throughout the region. The relatives of European lychnis moths that occur in North America do visit and pollinate—but do not lay eggs in—white campion flowers. New homes, however, aren't without new risks. A fungal disease called anther smut (*Microbotryum violaceum*) infects white campions both in Europe and North America, causing sterility; infected plants survive but cannot reproduce. The fungus produces spores that are shed from the anthers like pollen, and moths carry these spores from flower to flower—making anther smut a sexually transmitted disease much like chlamydia. The simple lovely fragrance and striking white color of white campions at dusk belies the incredible complexity of the economy of nature. The interactions between these plants and other organisms are a balance of costs and benefits.

9 PM
Sacred Datura
Datura wrightii

(NORTH AMERICA)

The white flowers of the sacred datura slowly unfurl as night falls in the desert. The buds are all twisted counterclockwise, the long tips of the petals overlapping one another in a tight spiral. Once open, these flowers are spectacular—up to twenty centimeters (eight inches) long and eleven centimeters (about four inches) in diameter, with a long and narrow tube, a broadly flaring mouth where the anthers protrude, and spreading petals with five whiplike tips. At the base of the tube is the nectar, that sugary solution sought out by the large hovering hawk moths who act as the primary pollinators of these nocturnal flowers. Hawk moths hover at the flower mouth like hummingbirds; this takes a lot of energy. And the sacred datura provides—the fifty to one hundred flowers on an individual plant can supply three to six thousand calories for these energy-hungry insects.

The flowers' white color stands out in the darkness, orienting the moths, while their odor acts as a long-range attractant. Moths perceive this scent with sensors on their antennae. As they approach, they also perceive the heightened carbon dioxide concentration that accompanies the flowers' opening. This close-range signal dissipates quickly, but the flowers produce another— and quite unexpected—one throughout the night: humidity. Plants transpire and exchange gases through tiny pores called stomata. As anyone who struggles to keep houseplants alive can tell you, the evaporation of moisture through stomata can cause leaves to wilt when plants don't have enough water. Moths are innately attracted to humid flowers and preferentially visit them even in the absence of a nectar reward. Aiming to please, the sacred datura actively transpires to elevate humidity levels inside its flowers.

The flowers maintain their humidity even when it's windy or a moth's wing flapping dries the air. For a plant growing in the desert, this comes at a cost. (Picture that sad, wilting houseplant.) But the benefits are too great to ignore. Experiments have shown that hawk moths ingest higher sugar concentrations and live longer when conditions are humid. Flower humidity, then, is an effective signal of a much-needed resource for the pollinators. The plant, in turn, benefits from increased flower visits from these insects.

People have also formed connections with the trumpet flowers of the sacred datura. In the cultures of the American Southwest, datura is one of the most powerful magic plants for those seeking to enter nonordinary reality. The mythical don Juan Matus, whose teachings were set down by Carlos Castaneda in the 1960s, considered sacred datura to be one of the "allies"—plants whose proper use conferred power. But datura, or *yerba del diablo* (devil's weed), could also be dangerous. Unpredictable and violent, if left uncontrolled, it destroyed. Ingesting sacred datura can produce the sensation of flying. As Castaneda wrote, "It is the root that gives direction and wisdom, and it is the cause of flying. As you learn more, and take it often in order to fly, you will begin to see everything with great clarity."

Like many plants in the nightshade family, including the coyote tobacco we met earlier, the sacred datura contains potent chemicals called tropane alkaloids. Nicotine is one such alkaloid, as is scopolamine, the "truth drug." These chemicals disrupt the normal signals sent to the human central nervous system, causing effects like the sensation of flying and changes in visual perception. Humans seeking these effects have used sacred datura as a recreational drug, sometimes with fatal consequences. But the ritual use of plant allies like the sacred datura involves more than just an experience; it is an interaction with purpose and meaning, not to be undertaken lightly.

10PM
Leafless Ephedra
Ephedra foeminea

(EUROPE)

In the pale light of the full moon in July, wiry ephedra plants in the scrubby vegetation of the eastern Mediterranean glitter as if sprinkled with diamonds. A closer look reveals these jewels to be reflections off the pollination droplets on the "flowers" of these shrubs.

Ephedras, which are related to pine trees and cypresses, do not really have flowers. Flowering plants are called angiosperms, meaning "hidden seeds"; their seeds are protected in an ovary that matures to form a fruit. The seeds of gymnosperms (from the Greek *gymnos*, meaning "naked") are usually held in cones without a protective covering. Gymnosperms are mostly dioecious, a bit like the white campions we met earlier: individuals are either pollen producing (staminate males) or ovule producing (pistillate females).

The wind pollinates most gymnosperms, wafting pollen from staminate trees so that it lands on the tiny droplets exuded from the tips of ovules in pistillate cones. Some gymnosperms rely on these droplets—fittingly called pollination droplets—for fertilization. It is through this water that pollen grains, using their flagella, swim to reach and fertilize the naked egg. Some gymnosperms, however, rely on insects for pollination, even though they lack the showy structures—flowers—that angiosperms use as attractants. Among these insect-reliant gymnosperms are close relatives of *Ephedra*: *Gnetum*, a tropical liana, and *Welwitschia*, a peculiar plant that grows only in the Namib Desert. Together these three genera form a group called Gnetales, whose fossil records reach back in time to the age of the dinosaurs.

In Gnetales the pollination droplets exuded from cones are not only a watery fluid through which the pollen reaches the ovules but also high in sucrose, a sugar that functions like nectar to attract insects. Botanists working in the eastern Mediterranean were long baffled by the huge variability in the timing of the insect-pollinated *Ephedra foeminea*'s reproduction; other, wind-pollinated species in the same area were reproductive at roughly the same time every year. Looking at their data, they realized that this ephedra produced pollination droplets in concert with the lunar cycle. Year after year, droplet production peaked on the day of the full moon in July. On these nights, tiny nocturnal moths and flies visit the ephedra cones.

Plants can detect the moon by the intensity of its light and gravity. Perhaps the combination of these two environmental factors allows the ephedra to track the lunar cycle. Insects, too, use moonlight—or else the angle between the moon and the horizon—as a navigational tool. The light of the full moon thus facilitates the pollination of *Ephedra foeminea* both by providing a navigational aid for nocturnal insects and by enhancing their attraction to the cones.

But this is not a highly tuned pollination system like many of those found in flowering plants. Many insects, from ants to flies and beetles to parasitic wasps, visit the colorful yellow-and-red cones of the ephedra during the day to feed on the sucrose-rich pollination droplets. It is not clear whether these too are pollinators or just visitors.

Today most other ephedras do not need insects for pollination, but this wasn't always the case. Fossil ephedra pollen from 125 million years ago is similar to that of our insect-pollinated ephedra species, suggesting that most ephedras relied on insects back then. Mass extinctions 60 million years ago reduced the diversity of all life, including ephedras and their relatives. When ephedra pollen again appears in the fossil record, it is like that of wind-pollinated species. It seems that ephedra switched from insect to wind pollination as ecosystems recovered, perhaps as a surer means of reproduction in fluctuating climates. This, in a story of evolution and survival, was its best chance at beginning to diversify again.

11 PM
Giant Saguaro
Carnegiea gigantea

(NORTH AMERICA)

The huge columnar saguaro cactus is a sentinel of the Sonoran Desert of northern Mexico and the southwestern United States. Growing as tall as twenty-three meters (about seventy-five feet), measuring three meters (about ten feet) or more in circumference, and weighing up to six tons, a saguaro can live for as long as two hundred years. It's only after many years of growth that they flower. These spectacular plants are sacred to the Tohono O'odham peoples, in whose lands they grow.

When a saguaro does bloom, the flowers only develop at the very top of the main column and branches, or arms. It looks a bit like the cactus is wearing white retro floral swim caps! Each saguaro flower is as large as eight centimeters (about three inches) in diameter, with many creamy-white petals and hundreds of pale-yellow anthers. At the flower base is nectar, as much as a teaspoon's worth in each flower—a rich resource for any animal able to access it. Between sunset and midnight in late April or early May, springtime in these northern deserts, the flowers of the saguaro open, each for a single night. These are classic bat flowers—large, sturdy, brilliantly white, and night blooming.

Nectar-feeding bats migrate from the south to the Sonoran Desert every spring, occupying nursery caves, where they raise their young. Chief among those that visit the saguaro flowers is the lesser long-nosed bat, whose elongate snout fits perfectly into the floral tube as the bat hovers at the flower mouth. The brushy tongue of these bats is ideal for lapping up nectar. To eat, they must push through the flower to reach the pool at its bottom, thereby covering their heads with pollen from the many anthers. Flowering lasts through June, coinciding with the time the migratory lesser long-nosed bats are resident in the Sonoran Desert. Competition for these effective pollinators among the columnar cacti, of which saguaro is the biggest, can be intense; the bats depend on nectar from these cacti but can be unreliable due to their low population densities away from nursery sites. But saguaro cacti have a strategy.

Saguaro flowers produce copious nectar at night, and while this reward is often lapped up by the bats, the flowers remain open until about noon the next day. This invites daytime visitors, who may be even better pollinators, to visit. Honeybees, introduced to the Americas by European colonizers, bustle around in saguaro flowers, collecting pollen and nectar. They contact the stigmas and transfer pollen pretty efficiently. But it is a third type of pollinator—the western white-winged dove, a migrant from the south—that really knows how to exploit these flowers. These birds nest and reproduce in the Sonoran Desert, and their numbers peak at the same time that the saguaro cacti flower. The doves are saguaro specialists, using the flowers and fruit as sources of both sugar and water. They insert their entire head into the flower to extract the nectar, getting covered with pollen in the process. Virtually all the doves in an area visit the flowers many times over, transferring pollen as they go.

Given how high these flowers are, how do humans—firmly planted on the ground—know what is going on and who is visiting? By using a giant saguaro selfie stick, of course. Strapping cameras to long poles, scientists and volunteers showed that the location of flowers on the tops of the columns changes throughout the blooming season. Flowers first open on the east-facing side of the cactus, and flowering gradually moves counterclockwise over the course of the season. Tracing the movements of the sun might be a protective strategy, allowing the flowers to avoid cold snaps early in the season and intense heat later in the summer. Or the cactus could be responding to the amount of sunlight available to make energy and speed up flower development.

Whatever mechanism is behind this phenomenon, the Tohono O'odham peoples certainly knew of the sequential flowering of the saguaro before it was "discovered" by scientists. Saguaro fruits and seeds are integral to the culture, and for there to be fruits and seeds, flowers and pollinators must come together.

Epilogue

We are now at the end of the day and at the end of our journey. Or are we? Flowers are everywhere, on every continent and in every habitat, and something is going on with them wherever and whenever they appear. We have met only a very few species here, but there are so many more. Flowering plants dominate the vegetation of most of Earth's ecosystems; they are the foundation of the animal world. The emergence of flowering plants some 150 to 250 million years ago kick-started a world of interactions—pollination, herbivory, and fruit and seed dispersal, to name a few. Scientists have argued about which came first, flowers or their pollinators, and which drives the diversity of the other. But I don't think that really matters. What matters is that the relationships exist at all and in all their wondrous forms. Whatever time of day you look for a flower—and as we have seen, they are there all day and night—look too for the other organisms. They will be there with it.

As we alter the planet to suit our own purposes, we often upset these intricate relationships. As we have seen, the timing of pollen presentation and stigma receptivity can have a big effect on cross-pollination and therefore on the genetic diversity needed for plants to adapt to a changing environment. Timing matters for blooming too: if flowers open too early, before the key pollinators have emerged, and no seed is set, no new generation of plants will be born. Evidence suggests that climate change is driving the increasing mismatch in the timing of blooming and pollinator abundance, as we have seen with the California poppies. Subtle differences in the timing of saguaro flowering and bat or dove migration could have similarly profound effects on the survival of these iconic plants. Plants are true survivors, though, and because many have mixed pollination strategies, they can often adapt. But we cannot rely on their capacity to absorb the wounds we inflict on the earth. As our own species tries to adapt to our rapidly changing world, we could learn a lot from plants and their flowers.

Whatever the time of day or night, watching a flower can reveal a world of interactions and wonders—some of which will have already been studied carefully and described by scientists, while others will be new. There is much still to be found out about flowers.

So look at a flower with fresh eyes and see what you can see. Georgia O'Keeffe was right when she said that a flower can be your whole world.

Acknowledgments

This book would never have been written without the kind and generous help of the many botanical colleagues who answered my endless questions about flowers and their lives. You know who you are, and many, many thanks! Choosing the twenty-four species was not easy—there are so many wonderful plants! Joe Calamia was instrumental in helping me focus the content and in carefully editing the final text. Joe, seeing flowers through your eyes made me see them in new ways. Katie Scott, thank you for finding the time to illustrate the day. Rob Knapp in Montana and Jim and Judy Knapp in Oregon provided the perfect places to write, with views of mountains and nature to be immersed in. Rob in particular endured many stories of flowers and their goings-on with great patience on a long and eventful road trip. The Natural History Museum in London is the perfect place in which to think creatively about how human lives intersect with nature, and I am fortunate to work there; it brings me joy every day. My love of flowers was and continues to be inspired by my family: my late parents, Ed and Jean Knapp, whose enthusiasm for the outdoors sparked mine; my siblings, Dave, Rob, and Mary, who endure many hikes and adventures with flowers; and my children, Alfred, Isabel, and Victor Mallet, who seem to have survived a childhood immersed in botany and emerged unscathed. And last but not least, thanks to the plants themselves—endlessly fascinating, endlessly changing. If you take the time, they'll let you into their lives.

Further Reading

PREFACE

Coen E (2001) Goethe and the ABC model of flower development. *Comptes Rendus de l'Académie des Sciences—Series III—Sciences de la Vie* 324: 523-30. Quoting Goethe JW (1970) *Italian journey*, trans. Auden WH, Mayer E (Penguin Classics, London, UK).

Goethe JW (1790) *Versuch die Metamorphose der Pflanzen zu erklären* [Metamorphosis of plants]. Gotha.

MIDNIGHT: MOONFLOWER

Galetto L, Bernardello G (2004) Floral nectaries, nectar production dynamics and chemical composition in six *Ipomoea* species (Convolvulaceae) in relation to pollinators. *Annals of Botany* 94: 269-80.

Hosler D, Burkett SL, Tarkanian MJ (1999) Prehistoric polymers: rubber processing in ancient Mesoamerica. *Science* 284: 1988-91.

1AM: QUEEN OF THE NIGHT

Darwin E (1789) *The botanic garden, part II: containing the loves of the plants*. London.

Linnaeus C (1753) *Species plantarum*. Stockholm.

2AM: ANGRAECUM

Hugel S, Michenau C, Fournel J, Warren BH, Gauvin-Bialecki A, Pailler T, Chase MW, Strasberg D (2010) *Glomeremus* species from the Mascarene Islands (Orthoptera, Gryllacrididae) with the description of the pollinator of an endemic orchid from the island of Réunion. *Zootaxa* 2545: 58–68.

Michenau C, Fournel J, Warren BH, Hugel S, Gauvin-Bialecki A, Pailler T, Strasberg D, Chase MW (2010) Orthoptera, a new order of pollinator. *Annals of Botany* 105: 355–64.

3AM: EAST COAST BANKSIA

Cunningham SA (1991) Experimental evidence for pollination of *Banksia* spp. by non-flying mammals. *Oecologia* 87(1): 86–90.

Hackett DJ, Goldingay RL (2001) Pollination of *Banksia* spp. by non-flying mammals in north-eastern New South Wales. *Australian Journal of Botany* 49: 637–44.

Wiens D, Rourke JP (1978) Rodent pollination in southern African *Protea* spp. *Nature* 276: 71–73.

4AM: CHICORY

Doyle T, Hawkes WLS, Massy R, Powney GD, Menz MHM,

Wotton KR (2020) Pollination by hoverflies in the Anthropocene. *Proceedings of the Royal Society B* 287: 20200508.

Gardiner BG (1987) Linnaeus' floral clock. *The Linnean* 3(1): 26–29.

Linnaeus C (1751) *Philosophia botanica*. Stockholm.

Perring FH, Walker KJ (2020) Chicory: *Chicorium intybus* L. *BSBI Online Plant Atlas 2020*, ed. Stroh PA, Humphrey TA, Burkmar RJ, Pescott OL, Roy DB, Walker KJ. Accessed June 5, 2023. https://plantatlas2020.org/atlas/2cd4p9h.xgb.

5AM: DANDELION

Frazee LJ, Morris-Marno S, Blake-Mahmud J, Struwe L (2016) Eat your weeds: wild and edible plants in urban environmental education and outreach. *Plant Science Bulletin* 62(2): 72–84.

Leguil S. More than Weeds. Accessed June 1, 2023. https://morethanweeds.co.uk/.

Richards AJ (2021) *Field handbook to British and Irish dandelions*. BSBI Handbook 23. Botanical Society of Britain and Ireland, Durham, UK.

Vickery R (2019) *Vickery's folk flora: an A–Z of the folklore and uses of British and Irish plants*. Weidenfield and Nicolson, London.

6AM: COYOTE TOBACCO

Haverkamp A, Hansson BS, Baldwin IT, Knaden M, Yon F (2018) Floral trait variations among wild tobacco populations influence the foraging behavior of hawkmoth pollinators. *Frontiers in Ecology and Evolution* 6: 19.

Kessler D, Baldwin IT (2008) Field experiments with transformed flowers reveal the sense of floral scents. *Science* 321: 1200-1202.

Kessler D, Diezel C, Baldwin IT (2010) Changing pollinators as a means of escaping herbivores. *Current Biology* 20: 237-42.

7AM: CURLY ROCK ROSE

Civeyrel L, Leclercq J, Demoly J-P, Agnan Y, Quèbre N, Pélissier C, Otto T (2011) Molecular systematics, character evolution, and pollen morphology of *Cistus* and *Halimium* (Cistaceae). *Plant Systematics and Evolution* 2965: 23-54.

8AM: UAPE JACANA OR GIANT WATERLILY

Prance GT, Arias JR (1975) A study of the floral biology of *Victoria amazonica* (Poepp.) Sowerby (Nympheaceae). *Acta Amazonica* 5(2): 109-39.

Smith LT, Magdalena C, Przelomska NAS, Pérez-Escobar OA, Melgar-Gómez DG, Beck S, Negrão R, Mian S, Leitch

IJ, Dodsworth S, Maurin O, Ribero-Guardia G, Salazar CD, Gutierrez-Sibauty G, Antonelli A, Monro AK (2022) Revised species delimitation in the giant water lily genus *Victoria* (Nymphaeaceae) confirms a new species and has implications for its conservation. *Frontiers in Plant Science* 13: 883151.

9AM: SUNFLOWER

Atamain HS, Creux NM, Brown EA, Garner AG, Blackman BK, Harmer SL (2016) Circadian regulation of sunflower heliotropism, floral orientation, and pollinator visits. *Science* 353: 587–90.

10AM: SAGO PALM

Chao S (2022) *In the shadow of the palms: more-than-human becomings in West Papua.* Duke University Press, Durham, NC.

Flach M (1997) *Sago palm*: Metroxylon sagu *Rottb.* Promoting the Conservation and Use of Underutilized and Neglected Crops 13. IPGCPR, Gaterslaben; IPGRI, Rome.

Schuiling DL, Jong FS, Flach M (1993) *Exploitation and natural variability of the sago palm: report of a Sarawak and all-Indonesia study tour, January–February 1992.* Agronomy Department, Tropical Crop Science Section, Wageningen Agricultural University, the Netherlands.

11AM: WESTERN BLUE FLAX OR LEWIS FLAX

Addicott FT (1977) Flower behavior in *Linum lewisii*: some ecological and physiological factors in opening and abscission of petals. *American Midland Naturalist* 97(2): 321-32.

Kearns CA, Inouye DW (1994) Fly pollination of *Linum lewisii* (Linaceae). *American Journal of Botany* 81(9): 1091-95.

Orford KA, Vaughan AP, Memmott J (2015) The forgotten flies: the importance of non-syrphid Diptera as pollinators. *Proceedings of the Royal Society B* 282: 20142934.

NOON: JACK-GO-TO-BED-AT-NOON OR MEADOW SALSIFY

Perring FH, Walker KJ (2020) Goat's-beard: *Tragopogon pratensis* L. *BSBI Online Plant Atlas 2020*, ed. Stroh P, Humphrey TA, Burkmar RJ, Pescott OL, Roy DB Walker KJ. Accessed June 5, 2023. https://plantatlas2020.org/atlas/2cd4p9h.ft7.

Soltis DE, Mavrodiev EF, Gitzendammer MA, Alexeev YE, Godden GT, Soltis PS (2022) *Tragopogon dubius*: multiple introductions to North America and the formation of the New World tetraploids. *Taxon* 71(6): 1287-98.

Soltis DE, Soltis PS, Pires JC, Kovarik A, Tate JA, Mavrodiev E (2004) Recent and recurrent polyploidy in *Tragopogon* (Asteraceae): cytogenetic, genomic and genetic comparisons. *Biological Journal of the Linnean Society* 82: 485-501.

1PM: ANTARCTIC HAIR GRASS

Cavieres LA, Sáez P, Sanhueza C, Sierra-Almeida A, Rabert C, Corcuera LJ, Alberdi M, Bravo LA (2016) Ecophysiological traits of Antarctic vascular plants: their importance in the responses to climate change. *Plant Ecology* 217: 343–58.

Fowbert JA, Lewis Smith RI (1994) Rapid population increase in native vascular plants in the Argentine Islands, Antarctic Peninsula. *Arctic and Polar Research* 26(3): 290–96.

González ML, Urdampilleta JD, Fasanella M, Premoli AC, Chiapella JO (2016) Distribution of rDNA and polyploidy in *Deschampsia antarctica* E. Desv. in Antarctic and Patagonic populations. *Polar Biology* 39: 1663–77.

Holtom A, Greene SW (1967) The growth and reproduction of Antarctic flowering plants. *Philosophical Transactions of the Royal Society B* 252: 323–37.

Parnikoza I, Dykyy I, Ivanets V, Kozeretska I, Kunakh V, Rozhok A, Ochyra R, Convey P (2012) Use of *Deschampsia antarctica* for nest building by the kelp gull in the Argentine Islands area (maritime Antarctica) and its possible role in plant dispersal. *Polar Biology* 35: 1753–58.

2PM: TITAN ARUM OR CORPSE FLOWER

Claudel C, Lev-Yadun S (2021) Odor polymorphism in deceptive

Amorphophallus species—a review. *Plant Signalling and Behaviour* 16(12): ee1991712.

Giordano C (1999) Observations on *Amorphophallus titanum* (Becc.) Becc. ex Archangeli in the forest of Sumatra. *Aroideana* 22: 10-19.

Korotkova N, Barthlott W (2009) On the thermogenesis of the Titan arum (*Amorphophallus titanum*). *Plant Signalling and Behaviour* 4(11): 1096-98.

Lobin W, Neumann M, Radscheit M, Barthlott W (1987) The cultivation of titan arum (*Amorphophallus titanum*)—a flagship species of botanical gardens. *Sibbaldia* 5: 69-86.

3PM: THE TRAVELER'S TREE

Birkinshaw CR, Colquhoun IC (1998) Pollination of *Ravenala madagascariensis* and *Parkia madagascariensis* by *Eulemur macaco* in Madagascar. *Folia Primatologia* 69: 252-59.

Haevermans T, Hladik A, Hladik C-M, Razanatsoa J, Haevermans A, Jeannoda V, Blanc P (2021) Description of five new species of the Madagascan flagship plant genus *Ravenala* (Strelitziaceae). *Scientific Reports* 11: 21965.

Kress WJ, Schatz GE, Andrianifahanana M, Morland HS (1994) Pollination of *Ravenala madagascariensis* (Strelitziaceae) by lemurs in Madagascar: evidence for an archaic coevolutionary system? *American Journal of Botany* 81(5): 542-51.

4PM: FOUR-O'CLOCK

Gandía-Herrero F, García-Carmona F, Escribano J (2005) Floral fluorescence effect. *Nature* 437: 334.

Suzuki M, Myahara T, Tokumoto H, Hakamatsuka T, Goda Y, Ozeki Y, Sasaki N (2014) Transposon-mediated mutation of CYP76AD3 affects betalain synthesis and produces variegated flowers in four o'clock (*Mirabilis jalapa*). *Journal of Plant Physiology* 171: 1586–90.

Zhang H, Zhao X, Zhao F, Han J, Sun K (2002) Mendel's controlled pollination experiments in *Mirabilis jalapa* confirmed his discovery of the gamete theory of inheritance in *Pisum*. *Hereditas* 159: 19. https://doi.org/10.1186/s41065-022-00232-1.

5PM: HARAKEKE OR NEW ZEALAND FLAX

Craig JL, Stewart AM (1988) Reproductive biology of *Phormium tenax*: a honeyeater-pollinated species. *New Zealand Journal of Botany* 26: 453–63.

Hindmarsh G (1999) Flax—the enduring fibre. *New Zealand Geographic* 42. https://www.nzgeo.com/stories/flax-the-enduring-fibre/.

Hughes R (1986) *The fatal shore: a history of the transportation of convicts to Australia, 1787–1868*. Vintage Books, London.

6PM: CALIFORNIA POPPY

Jordan DS (1922) The California poppy. *Science* 56: 168–69.

Pearson KD, Love NLR, Ramirez-Parada T, Mazer SJ, Yost JM (2021) Phenological trends in the California poppy (*Eschscholzia californica*): digitized herbarium specimens reveal intraspecific variation in the sensitivity of flower date to climate change. *Madroño* 68(4): 343–49.

Wilts BD, Rudall PJ, Mayroud E, Gregory T, Osawa Y, Vignolini S, Steiner U, Glover BJ (2018) Ultrastructure and optics of the prism-like petal epidermal cells of *Eschscholzia californica* (California poppy). *New Phytologist* 219: 1124–33.

7PM: NIGHT-BLOOMING JESSAMINE

Heath RR, Landolt PJ, Dueben B, Lenczewski B (1992) Identification of floral compounds of night-blooming jessamine attractive to cabbage looper moths. *Environmental Entomology* 21(4): 854–59.

Overland L (1960) Endogenous rhythm in opening and odor of flowers of *Cestrum nocturnum*. *American Journal of Botany* 47(5): 378–82.

Patil CD, Patil SV, Salunke BK, Salunkhe RB (2011) Bioefficacy of *Plumbago zeylanica* (Plumbaginaceae) and *Cestrum nocturnum* (Solanaceae) plant extracts against *Aedes aegypti*

(Diptera: Culicidae) and nontarget fish *Poecilia reticulata*. *Parasitology Research* 108: 1253-63.

8PM: WHITE CAMPION

Bernasconi G, Antonovics J, Biere A, Charlesworth D, Delph LF, Filatov D, Giraud T, Hood ME, Marais GAB, McCauley D, Pannell JR, Shykoff JA, Vyskot B, Wolfe LM, Widner A (2009) *Silene* as a model system in ecology and evolution. *Heredity* 103: 5-14.

Dotterl S, Jürgens A, Seifert K, Laube T, Weissbecker B, Schütz S (2006) Nursery pollination by a moth in *Silene latifolia*: the role of odours in eliciting antennal and behavioural responses. *New Phytologist* 169: 707-18.

Labouche A-M, Bernasconi G (2010) Male moths provide pollination benefits in the *Silene latifolia-Hadena bicruris* nursery pollination system. *Functional Ecology* 24: 534-44.

Young HJ (2002) Diurnal and nocturnal pollination of *Silene alba* (Caryophyllaceae). *American Journal of Botany* 89(3): 433-40.

9PM: SACRED DATURA

Bronstein JL, Huxman T, Horvath B, Farabee M, Davidowitz G (2009) Reproductive biology of *Datura wrightii*: the benefits of a herbivorous pollinator. *Annals of Botany* 103: 1435-43.

Casteneda C (1969) *The teachings of Don Juan: a Yaqui way of knowledge*. University of California Press, Berkeley.

Dahake A, Jain P, Vogt CC, Kandalaft W, Stroock AD, Raguso RA (2023) A signal-like role for floral humidity in a nocturnal pollination system. *Nature Communications* 13: 7773.

Raguso RA, Henzel C, Buchmann SL, Nabhan GP (2003) Trumpet flowers of the Sonoran Desert: floral biology of *Peniocereus* cacti and sacred *Datura*. *International Journal of Plant Sciences* 164(6): 877–92.

10PM: LEAFLESS EPHEDRA

Bolinder K, Humphreys AM, Ehrlén J, Alexandersson B, Ickert-Bond SM, Rydin C (2016) From near extinction to diversification by means of a shift in pollination mechanism in the gymnosperm relict *Ephedra* (Ephedraceae, Gnetales). *Botanical Journal of the Linnean Society* 180: 461–77.

Di Stilio VS, Ickert-Bond SM (2021) *Ephedra* as a gymnosperm evo-devo model lineage. *Evolution and Development* 23: 256–66.

Rydin C, Bolinder K (2015) Moonlight pollination in the gymnosperm *Ephedra* (Gnetales). *Biology Letters* 11: 21040993.

11PM: GIANT SAGUARO

Fleming TH, Sahley CT, Holland JN, Nason JD, Hamrick JL

(2001) Sonoran Desert columnar cacti and the evolution of generalized pollination systems. *Ecological Monographs* 71(4): 511–30.

Foley T, Swann DE, Sotelo G, Perkins N, Winkler DE (2021) Asynchronous flowering patterns in saguaro cacti (*Carnegiea gigantea*). *Ecosphere* 12(2): e03873.

McGregor SE, Alcorn SM, Olin G (1962) Pollination and pollinating agents of the saguaro. *Ecology* 43(2): 259–67.

Wolf BO, Martinez del Rio C (2000) Use of saguaro fruit by white-winged doves: isotopic evidence of a tight ecological association. *Oecologia* 124: 536–43.

Index

PAGE NUMBERS IN ITALICS REFER TO ILLUSTRATIONS.

Acacia spp. (wattles), 22
Africa, 47. *See also* Mauritius; South Africa
Amazon, xvii, xix, 53-58
Americas, 3, 5, 103; honeybees introduced to, 152. *See* Central America; Mesoamerica; North America; South America
Amorphophallus titanum. See corpse flower
angiosperms, xiii, 143-45
angraecum, 14, *15*-19
Angraecum cadetii. See angraecum
Angraecum sesquipedale, 15-16
animals, xv-xvi, 23, 43-44, 92, 125, 131, 150, 155; as pollinators, xviii-xix. *See also* mammals; *and specific animals*
Antarctica, 83-88
Antarctic hair grass, xiii-xiv, xviii, 83-88, *84*
anthers, xii, xiv, xvii-xviii, 11, 29, 44, 74-75, 79, 86, 99-100, 113-14, 115, 129, 134, 137, 150-51. *See also* pollen
anther smut (*Microbotryum violaceum*), 134
ants, 146-47
Aotearoa (New Zealand), 109-14; endemic species on decline in, 114
apomixis, xiv-xv
Asia. *See* Eurasia; Southeast Asia
Asteraceae, 29
Australia, 21-25, 111-12
autumn, 33
Aztecs, 104

Banksia integrifolia. See east coast banksia
banksias. *See* east coast banksia
bats, xviii-xix, 13, 100, 150-52, 157
bees, xi-xii, xviii-xix, 31-32, 43-44, 50, 74-76, 93-94, 103, 115, 117. *See also* honeybees
beetles, 53, 55-57, 93-94, 146-47

beets, 105
benzyl acetone, 41–42
benzyl alcohol, 125
betalains, 105
biology. *See* reproduction
birds, xi–xii, xviii, xix, 17–18, 21, 23, 44, 103, 112–14, 152. *See also specific birds*
birds-of-paradise, 98
blue flax. *See* western blue flax
botany, 11–12, 58, 74, 80, 160
bougainvilleas, 105
bumblebees, 76

cabbages, 68, 126
cacti: betalains of, 105; night-flowering, xi; pollinators of, 151; tepals of, 11. *See also* giant saguaro; queen of the night; saguaro cacti
Cactus grandiflorus, 12
California poppy, xix, 115–21, 116, 157; as *dormidera* (sleepy one), 120
carbon dioxide, 138
Caribbean, 9, 123–24
carnations, 105
Carnegiea gigantea. See giant saguaro
Castaneda, Carlos, 140–41

caterpillars, 41–42, 44, 126, 132–33
Central America, 9
Cervantes, Miguel de, 47, 52
Cestrum nocturnum. See night-blooming jessamine
chicory, 26, 27–32, 37, 59, 79
Cichorium endivia, 30
Cichorium intybus. See chicory
circadian rhythm, 61, 63, 124
Cistus albidus, 50–51
Cistus crispus. See curly rock rose
Cistus species, and hybridization, 49–52. *See also* curly rock rose
climate change, 88, 118, 120, 147, 156–57
cloves, 125
common flax (*Linum usitatissimum*), 73
Compositae, 29
composites, 29, 37, 59, 79
Cook, Captain James, 111
corpse flower, 89–95, 90
COVID-19, 58
coyote tobacco, 39–45, 40, 141
crickets, xix, 17–19
cross-fertilization, 114
cross-pollination, 51, 57, 156
curly rock rose, 46, 47–52, 81
cypresses, 143

daisies, 29
dandelion, xiv, 33–38, *34*, 59, 79
Darwin, Charles, 16
Darwin, Erasmus, 10–11, 13
Datura wrightii. *See* sacred datura
daytime pollinators, 42, 99–100, 152
Deschampsia antarctica. *See* Antarctic hair grass
deserts, 137, 139. *See also* Namib Desert; Sonoran Desert
devil's weed (*yerba del diablo*). *See* sacred datura
dinosaurs, 145
dioecious plants, xv–xvi, xviii, 131, 144
doves, 152–53, 157

east coast banksia, xvii, *20*, 21–25
ecosystems, 105–6, 147, 155. *See also* habitats
endangered species, 101, 128
Ephedra foeminea. *See* leafless ephedra
Eschscholzia californica. *See* California poppy
eugenol, 125
Eurasia, 129
Europe, 5–6, 11–12, 27, 47, 73, 77, 80–81, 109–11, 114, 132, 134, 143; honeybees introduced to Americas from, 152; wildflowers of, 31
evolution: and plant genomes, 81; and solar tracking, 64; and survival, 147

fertilization: cross-, 114; and pollination droplets, 144; self-, xvi–xviii, 23, 49, 74, 131–32. *See also* reproduction
flax. *See* common flax (*Linum usitatissimum*); harakeke; western blue flax
flies, xix, 74–76, 93–94, 146–47. *See also* hoverflies
florets, 37–38, 85–86
flowerflies. *See* hoverflies
flowers and plants: adaptability, resilience, and survival of, xxi, 157; emergence of, 155–56; as endlessly changing, 160; as foundation of animal world, 155; fragility of, 13; interactions of, 135, 157; power and beauty of, xxi; presence of, 155. *See also specific flowers*
forests, 10, 17, 56, 65, 67–70, 93, 97–98, 104–5. *See also* rainforests
fossils, 145, 147
four-o'clock, xi, 102, 103–7

fruits, 95, 119, 126, 133, 143-44, 152; and flowering, 65, 67; and seeds, xii, 154-55

fungal diseases, 134

gallant of the night (*galán de noche*), 124

gardens: botanical, 10, 58, 89, 93; *Cistus* species in, 49; public, xx; tropical, 128. *See also* greenhouses

Gebze, Marcus, 70

genetics, xvii, 35, 43, 57, 75, 80-81, 87, 105-6, 114, 131, 156

germination, 61-62

giant saguaro, *148*, 149-54, 157; as sacred, 149

giant waterlily. *See* uape jacana

gingers, 98

global warming, 88. *See also* climate change

Glomeremus orchidophilus, 18

Gnetales, 145

Gnetum, 145

Goethe, Johann Wolfgang von, xx

greenhouses, 11, 57-58, 128. *See also* gardens

Guianas, 53-58

gymnosperms, 144-45. *See also* angiosperms

habitats, 51-52, 118, 155. *See also* ecosystems

hair grass. *See* Antarctic hair grass

harakeke, *ii*, *108*, 109-14; as protandrous, 113-14

Hawaii, 128

hawk moths, 4-5, 16, 41-44, 103, 137-39

heart of palm, 68

Helianthus annuus. *See* sunflower

heliotropism, 62. *See also* phototropism

herbivory, 155

honeybees, 152

hoverflies, 31-32, 75. *See also* flies

humidity, 138-40

hummingbirds, 4, 42-43, 103, 138

hybridization, 49-52, 80-81

inflorescences, 21-24, 31, 37-38, 65, 67, 79, 85, 91-92, 98-100, 110, 112-14

infrared radiation, 94-95

insects, xix, 4, 15, 17-18, 21, 23, 29, 31, 37, 41, 74-76, 93-95, 126, 138, 140, 144-47. *See also specific insects*

invasive species, 114, 134

Ipomoea alba. *See* moonflower

iridescence, 48-49, 120

Jack-go-to-bed-at-noon, 77-82, 78
jasmine, 123, 125
jessamine. *See* night-blooming jessamine

leafless ephedra, 142, 143-47
lemurs, xi-xii, xix, 97, 99-101
lettuce, 30, 37
Lewis, Meriwether, 73-74
Lewis flax. *See* western blue flax
Linnaeus, Carolus, 12, 27-29, 32
Linum lewisii. *See* western blue flax
Linum usitatissimum (common flax), 73
looper moths, 126
lunar cycle, 145-46

Madagascar, 15-19, 97-101, 110
Malagasy language, 98
Malagasy people, 100
mammals, xix, 23-25. *See also specific mammals*
Māori, 109-10, 112
Marind people, 69
marsupials, 21-23, 25
Marvell, Andrew, 28
Mascarene Islands, 16-18

mass extinctions, and diversity of life, 147
Mauritius, 17-19
meadow salsify. *See* Jack-go-to-bed-at-noon
Mediterranean, 48, 52, 143, 145
Mendel, Gregor, 106-7
Mesoamerica, 5-6
Metroxylon sagu. *See* sago palm
Mexico, 6, 9, 104, 149
Microbotryum violaceum (anther smut), 134
microspecies, 35-36
midsummer, 27, 83. *See also* summer
Mirabilis jalapa. *See* four-o'clock
monocarpy, 67
monocultures, 69-70
monoecious plants, xv. *See also* dioecious plants
moon: full, 143, 145-46; gravity, 146; moonlight, 13, 143, 146
moonflower, 2, 3-7; and rubber production, 5-6
morning glories, 3. *See also* moonflower
mosquitoes, 127
moths, 4-5, 13, 16-17, 103, 125-27, 132-34, 138-39, 146. *See also* hawk moths

mustards, 126

Nägeli, Carl Wilhelm von, 106-7
Namib Desert, 145
natural selection, and genetic variation, xvi-xvii
nectar, xix, 4, 13, 16-18, 23-25, 31-32, 41-44, 74, 97, 99-101, 103, 113-14, 125, 132-33, 137-39, 145, 150-53; and pollen, 21, 32, 37, 99-100, 152-53
New Guinea, 65
New Zealand (Aotearoa), 109-14; endemic species on decline in, 114
New Zealand flax. *See* harakeke
Nicotiana attenuata. *See* coyote tobacco
nicotine, 43-44, 141. *See also* coyote tobacco
night-blooming jessamine, xvi, 122, 123-28, 131-32
night-flowering species, xi, 12, 39, 103, 150
nightshade, 126, 141
nighttime pollinators, 13, 21-22, 24-25, 41, 55-56, 132, 146, 150-51
North America, 5-6, 35, 39, 59, 71, 80, 115, 134, 137, 149; species introductions to, 27, 77, 129, 132

O'Keeffe, Georgia, xi, 158
orchids, 15-18, 75. *See also* angraecum
outcrossing, and genetic variation, xvii-xviii
ovules, and pollen, xii-xvi, *xiv*, xviii, xx, 18, 99-100, 106, 131, 144

palms. *See* sago palm
Palouse region (Washington state), 80
Patagonia, 83-88
Peru, 104-5
pesticides, 126
petals, xi-xii, *xiv*, 11, 22, 29, 37, 47-49, 55-56, 59, 71, 73-74, 76, 86, 92, 99, 104, 114, 115, 119-20, 124-25, 129, 137, 150. *See also* tepals
Phormium tenax. *See* harakeke
phototropism, 61-62
pine trees, 143
pistils, xiii, *xiv*, xv-xvi, 55, 92, 131
plants. *See* flowers and plants
poisonous species, 126-28
pollen, xi-xviii, 11, 21-23, 25, 31-32, 55, 74-75, 80-81, 86, 92-94, 99-101, 113-14, 117, 129, 133-34, 144-45, 147, 151-53, 156; and

nectar, 21, 32, 37, 99–100, 152–53; and ovules, xii–xvi, xviii, xx, 18, 99–100, 106, 131. *See also* anthers; pollination

pollination, xi–xii, xviii–xxi, 17–18, 21, 24–25, 31–32, 41–44, 49, 63–64, 74–76, 93–94, 105–7, 112–14, 117–20, 131–34, 143–47, 152, 154–57; animals as pollinators, xviii–xix; cross-, 51, 57, 156; and genetic diversity, 156; and humidity, 139–40; and interactions, 107; self-, 114; in wild, 93. *See also* daytime pollinators; nighttime pollinators; pollen

Popol Vuh (K'iche' Mayan creation myth), 6–7

poppies. *See* California poppy

predators, 44

protandry, 113–14

Proteus, xx–xxi

Public Gardens of Palermo, xx

queen of the night, xviii–xix, 8, 9–13, 124

rain, 48, 117–18

rainforests, xvi, 93

Ravenala agatheae, 101

Ravenala madagascariensis. *See* traveler's tree

reproduction, xii–xviii, *xiv*, xx, 35, 64, 86, 92, 118, 129, 131–33, 144–45, 147. *See also* fertilization

research, and collaboration, 107

Réunion, 17–18

rock roses. *See* curly rock rose

Rocky Mountains, 71, 75

roses. *See* curly rock rose

Royal Botanic Gardens, Kew (London), 58

sacred datura, 136, 137–41; as "allies," 140–41; as dangerous, 140–41; magical power of, 140–41; as recreational drug, 141; ritual use of, 141; trumpet flowers of, 140

sago palm, 65–70, 66

saguaro cacti, xix, 149, 151–52, 157. *See also* giant saguaro

salsify. *See* Jack-go-to-bed-at-noon; *Tragopogon dubius* (yellow salsify)

saponins, 127

scientific research, and collaboration, 107

scopolamine, 141

seed dispersal, 81, 87, 155

seeds, xii-xiii, xviii, 76, 131; and fruits, xii, 154-55
Selenicereus grandiflorus. See queen of the night
sepals, xi, *xiv*, 48, 92, 119. *See also* petals; tepals
sesquiterpenes, 36
sessile plants, 39
Silene latifolia. See white campion
solar tracking, 62-64, 153
Sonoran Desert, 149-52
South Africa, 24
South America, 87, 104-5
Southeast Asia, 65
Southwest (United States), 140, 149
Spain, 6, 47-48, 50, 52, 120
spathes, 91-92, 94-95
species, micro-, 35-36
spring, xviii, 33, 37, 47, 117-18, 150
stamens, xii, *xiv*, xv-xvi, 22, 49, 55-56, 74, 86, 92, 131
stomata, 138-39
Sumatra, 89-95
summer, 27, 39, 85, 88, 153. *See also* midsummer
sun. *See* solar tracking; sunlight
sunbirds, 100
sunflower, 29, 59-64, *60*, 79, 124
sunlight, 57, 83, 128, 153
superblooms, 117, 120-21

superflower structures, 29, 59
syrphids. *See* hoverflies

Taraxacum officinale. See dandelion
tepals, xi, 11-13. *See also* petals; sepals
titan arum. *See* corpse flower
tobacco. *See* coyote tobacco; nicotine
Tohono O'odham peoples, 149, 154
Tragopogon dubius (yellow salsify), 80
Tragopogon miscellus, 80-81
Tragopogon pratensis. See Jack-go-to-bed-at-noon
traveler's tree, xv, xix, 96, 97-101, 110
Trigona (stingless bees), 93-94
tropane alkaloids, 141
tropical areas, xvi, 5, 57, 128, 145. *See also* rainforests
truth drug, 141
tubers, xiv, 91

uape jacana, xvii, xix, 53-58, *54*

vegetables, 89
Victoria amazonica. See uape jacana

Victoria boliviana, 58
Victoria cruziana, 58

Washington state, 80
wasps, xix, 15, 31, 44, 146–47
water lilies. *See* uape jacana
wattles (*Acacia* spp.), 22
weeds: and diversity, creation of, 81–82; and land-use practices, 82. *See also* dandelion; Jack-go-to-bed-at-noon
Welwitschia, 145
western blue flax, xv, 71–76, 72
West Papua (Indonesia), 67–70

white campion, xv, 129–35, 130, 144; and fungal disease, 134
wildflowers, European, 31. *See also* California poppy
winter, 117

yellow salsify (*Tragopogon dubius*), 80
yerba del diablo (devil's weed). *See* sacred datura

Zahara de los Atunes, Spain, 47–48, 50, 52

EARTH DAY

ALSO IN THE EARTH DAY SERIES

Bird Day:
A Story of 24 Hours and 24 Avian Lives
WRITTEN BY MARK E. HAUBER AND
ILLUSTRATED BY TONY ANGELL

Frog Day:
A Story of 24 Hours and 24 Amphibian Lives
WRITTEN BY MARTY CRUMP AND
ILLUSTRATED BY TONY ANGELL

Mushroom Day:
A Story of 24 Hours and 24 Fungal Lives
WRITTEN BY ALISON POULIOT AND
ILLUSTRATED BY STUART PATIENCE

Tree Day:
A Story of 24 Hours and 24 Arboreal Lives
WRITTEN BY MEG LOWMAN AND
ILLUSTRATED BY THIBAUD HÉREM